Beyond Heredity and Environment

Myrtle McGraw and the Maturation Controversy

edited by

THOMAS C. DALTON
AND VICTOR W. BERGENN

Westview Press

BOULDER • SAN FRANCISCO • OXFORD

Frontispiece shows Myrtle McGraw holding infant in her laboratory at Briarcliff College, 1969 (courtesy of Francis Mechner).

Copyright © 1995 by Westview Press, Inc.

Published in 1995 in the United States of America by Westview Press, Inc., 5500 Central Avenue, Boulder, Colorado 80301-2877, and in the United Kingdom by Westview Press, 12 Hid's Copse Road, Cumnor Hill, Oxford OX2 9JJ

Library of Congress Cataloging-in-Publication Data
Beyond heredity and environment : Myrtle McGraw and the maturation
  controversy / edited by Thomas C. Dalton and Victor W. Bergenn.
     p.   cm.
  Includes bibliographical references and index.
  ISBN 0-8133-2153-0
  1. Nature and nurture.   2. McGraw, Myrtle B. (Myrtle Byram),
1899—Contributions in child development.   3. Maturation
(Psychology)   4. Child development.   5. Developmental psychobiology.
I. Dalton, Thomas Carlyle.   II. Bergenn, Victor W.
BF341.B48   1995
150'.92—dc20   94-24018
                                                                                        CIP

Printed and bound in the United States of America

The paper used in this publication meets the requirements
of the American National Standard for Permanence of Paper
for Printed Library Materials Z39.48-1984.

10   9   8   7   6   5   4   3

Beyond Heredity and Environment

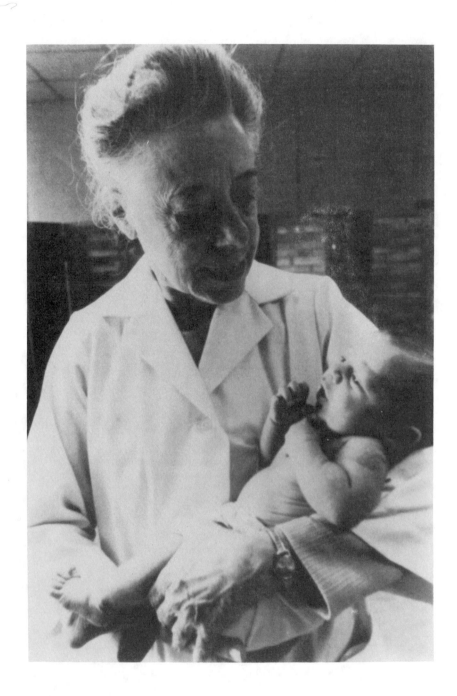

Contents

PART THREE
COGHILL, NEUROEMBRYOLOGY, AND THE
PRINCIPLES OF DEVELOPMENT

PART FOUR
PSYCHOBIOLOGY AND THE INTERDISCIPLINARY
STUDY OF INFANT DEVELOPMENT

.

Foreword: Myrtle McGraw's Nascent and Pioneering Use of Embryology to Understand Human Development

RONALD W. OPPENHEIM

The history of the study of behavioral development in this century has been marked by empirical controversies, recurring cycles of theoretical debates, misconceptions, personality cults, and occasionally—but only occasionally—by thoughtfulness, increased understanding of the issues, and genuine progress (Oppenheim 1982a; 1992). The 10-year-span between 1930 and 1940, when the lion's share of empirical and conceptual contributions of Myrtle McGraw took place, was a period of unprecedented intensive study of behavioral development that occurred within the context of heated theoretical and empirical controversies and debates. Indeed, in large measure, the controversy and debates drove the empirical studies of this era. Instinct, learning, environment, heredity, genes, maturation, reflexes, integration, and individuation are only a few of the key ideas and concepts of the period that were matters of vigorous discussion. For the most part, psychologists, psychiatrists, and pediatricians of the period who were interested in the behavioral development of animal and human infants carried out their investigations within the framework of behaviorism, in which experience, use, practice, learning, and conditioning were thought to be the major, if not the sole, mechanisms or processes driving the ontogeny of behavior.

In embryology, which was the major *biological* discipline concerned with developmental problems, the period from around the turn of the century to 1940 was a halcyon era during which the centuries old debate between preformation and epigenesis was finally settled amidst a plethora of fundamentally important advances in our understanding of cell biology, genetics, and embryogenesis. The triumph of epigenesis over preformation was especially noteworthy in that it provided a valid conceptual framework for guiding developmental studies of all kinds, including studies of neurobehavioral and psychological development (Oppenheim 1982a). Moreover, epigenesis provided a means for resolving the old, but still (at the time) nagging debates over nature *versus* nurture, heredity *versus* environment, and maturation *versus* learning. E.B. Wilson, the preeminent biologist of the time, summarized the new view by commenting that

> every living organism at every stage of its existence reacts to its environment by physiological and morphological changes. The developing embryo, like the adult, is a moving equilibrium—a product of the response of the inherited organization to the external stimuli working upon it. If these stimuli be altered, development is altered—but we cannot regard specific forms of development as directly caused by external conditions for the character of the response is determined not by the stimulus but by the inherited organization (Wilson 1900:428; 430).

Accordingly, by 1900 a somewhat sketchy and incomplete, but nonetheless conceptually valid understanding of development was available which has served as a foundation for all subsequent progress in this century.

Although early embryologists were mainly concerned with cellular and morphological development, the problem of behavioral development didn't escape their attention (Oppenheim 1982a; 1982b; 1992). Neuroembryologist Paul Weiss noted, "No account of neurogenesis can be complete without relating itself to the problem of behavior" (1955: 390). Many of the most prominent embryologists of this century including Ross Harrison, Steven Detwiler, Paul Weiss, Roger Sperry, V. Twitty, Rita Levi-Montalcini and Victor Hamburger conducted behavioral studies of embryos. However, one embryologist of this era stands above all the others in this regard and that is George E. Coghill. Beginning in 1906 and continuing until his death in 1941, Coghill (1929) made detailed observations of the behavioral and neuroanatomical development of salamander embryos and larvae. As a result of his studies, it became clear that the development of salamander behavior was related to orderly, predictable, and relatively stereotyped changes in the anatomy and physiology of the nervous system. From his observations of behavioral development in salamanders, Coghill also concluded that behavior is integrated as a total pattern from the beginning, and only later, and secondarily, do individual or independent activities emerge out of the total pattern. This idea, known as the "principle of the integration and individuation of behavior," was the major guiding force in all of Coghill's later thoughts on neurobehavioral development (Oppenheim 1978).

The approach of Coghill, as well as that of other embryologists, to the study of behavioral development differed fundamentally from that of psychologists and behaviorists (Oppenheim and Haverkamp 1985). First, embryological studies of behavior were carried out in the framework of biological epigenesis, whereas psychologists were more influenced by theories of learning and conditioning. Second, embryologists contended that the basic anatomical organization of the nervous system arises prior to the onset of behavioral function and is a major factor in determining specific behavior patterns, whereas psychologists believed that early function and behavior played a major role in establishing the structure of the nervous system. Finally, embryologists were concerned with the biological significance (adaptiveness) of developing behaviors, whereas psychologists tended to focus on developing behavior as an antecedent to adult behavior (Oppenheim 1981; Hall and Oppenheim 1987). Because of these differences, there was little understanding (and even less interaction) between the adherents of these two approaches to behavioral development. Psychologists felt that embryologists attributed everything in development to nature, maturation, and heredity, whereas embryologists characterized the psychological approach as an attempt to explain all of development by reference to nurture, learning, and environment. This was the situation that existed in 1930 when McGraw began her classic studies of behavioral development in human infants.

McGraw's investigations were conducted at Columbia University Babies Hospital as part of a correlated neural-behavioral examination of development in human infants under the direction of the neurologist Frederick Tilney. In keeping with the zeitgeist (see above) the approach of these studies was Coghillian in that an attempt was made to relate behavioral development to underlying neurobiological changes. In addition to Tilney, an advisory group of distinguished scientists and colleagues, including Coghill, John Dewey, John B. Watson, Robert Woodworth, as well as a number of pediatricians, counseled McGraw regarding the conduct and interpretation of her studies. The work was generously supported by the Rockefeller Foundation. Because there were considerable efforts underway throughout the United States at this time to study infant development, it is reasonable to ask what it was about this particular program of study that even now, over 60 years later, draws attention and praise to McGraw's investigations.

In my opinion, there are several factors that help account for the uniqueness and enduring nature of her work. First, was the attempt to understand human behavioral development by reference to changes in the central nervous system. Although at the time, other studies of this sort were being conducted on prenatal and postnatal animals, and on human fetuses, this was the first attempt to do so for human infants. Second, the methodology and the empirical approach differed from that used by most other investigators of human behavioral development, including the contemporary work of the much better known pediatrician, Arnold Gesell, at Yale University. Whereas Gesell used scales, landmarks, and standardized exams, McGraw watched what babies actually did and then devised ingenious means for manipulating their behavior. Third, whereas most investigators were primarily concerned with cataloging specific hallmarks of development (e.g., when a behavior such as standing-alone occurs), she was more concerned with the process of development (i.e., how and why behavior changes). Fourth, more than any

First Symposium for Growth and Development, North Truro, Massachusetts, 1939. *First row (left to right)*: Drs. L.C. Dunn, Columbia; V. Hamburger, Washington University; W.H. Lewis, Carnegie Institution; M.G. Brown, Washington University; R.L. Risley, State University of Iowa; L. Walp, Marietta College, Ohio; C.L. Schneider, Harvard; R.S. Childs, Columbia; E. Brill, Harvard; R. Gillette, Washington University. *Second row (left to right)*: Drs. L. Loeb, Washington University; W.F. Dove, University of Maine; T.B. Steele, Lankenau Hospital Research Institute; Mrs. L.G. Barth; Mrs. J. Needham; Dr. F. Peebles, Chapman College; Mrs. A.B. Dawson; Dr. F.S. Hammett, Lankenau Hospital Research Institute; Mrs. L. Loeb; K. Hyde, Lankenau Hospital Research Institute; Drs. L.P. Wilson, Wellesley; J. Outhouse, University of Illinois; M.B. McGraw, Columbia (indicated by the arrow); Mrs. V. Dammann, Columbia. *Third row (left to right)*: Drs. P.W. Gregory, University of California, Davis; C. Deuber, Yale; H. Dorsey, University of Connecticut; P. White, Rockefeller Institute; E.W. Sinnott, Columbia; P.A. Weiss, University of Chicago; A.B. Dawson, Harvard; R. Grant, McGill University; O. Rahn, Cornell; R. Aronson, Lankenau Hospital Research Institute; H.S. Burr, Yale; C.H. Waddington, Cambridge University; N.J. Berrill and G.H. Sander, McGill University. *Fourth row (left to right)*: Drs. J.W. Wilson, Brown University; L.B. Clark, Union College; L.G. Barth, Columbia; K.V. Thimann, and L. Hoadley, Harvard; A.D. Mead, Brown University; N. Padis and S.P. Reimann, Lankenau Hospital Research Institute; J. Needham, Cambridge University; O. Glaser, Amherst; G. Toennies and B. Miller, Lankenau Hospital Research Institute; J.F. Daniel, University of California, Berkeley; O. Schottee, Amherst; J.H. Woodger, University of London; C. Stern, University of Rochester; G. Smith, Yale. (Photo courtesy of Paul A. Weiss and Jane M. Oppenheimer.)

of her contemporaries (including Gesell), McGraw was firmly committed to carrying out and interpreting her work within the framework of experimental embryology. She was committed to the belief, as she put it, that "it is the experimental embryologists and not the psychologists who deserve credit for formulating the most adequate theory of behavioral development..." (McGraw 1935:10). Indeed, it was this commitment that led her to participate (as the only psychologist) in the first Growth symposium held in Massachusetts in 1939 together with embryologists and biologists of the time at which a broad range of developmental issues were discussed. The proceedings were published as *Growth, A Journal for Studies of Development and Increase* in 1939. In my opinion, it was also her recognition of the value of the methods and principles of experimental embryology, and not simply the desire to test the nature-nurture issue, that led her to undertake the perturbation studies with Johnny and Jimmy. To quote from her 1935 book, "Just as the embryologist may undertake to ascertain the effect of variation in temperature, light, and moisture upon the growth of particular organisms, we were interested in the influence of exercise or use of an activity upon its rate and mode of development" (McGraw 1935:24). Finally, in addition to the above mentioned factors, I would be remiss if I didn't also include here what may have been the most important factor of all in the consistent recognition and enduring nature of her work over the decades. Namely, the unique mind and personality that was Myrtle McGraw.

References

Coghill, G.E. 1929. *Anatomy and the Problem of Behavior.* Cambridge, UK: Cambridge University Press.

Hall, W.G. and R.W. Oppenheim. 1987. "Developmental Psychobiology: Prenatal, Perinatal, and Early Postnatal Aspects of Behavioral Development." *Annual Review of Psychology* 38:91–128.

McGraw, M.B. 1935. *Growth: A Study of Johnny and Jimmy.* New York: Appleton Century Crofts.

Oppenheim, R.W. 1978. "G.E. Coghill (1872–1941): Pioneer Neuroembryologist and Developmental Psychobiologist." *Perspectives in Biology and Medicine* 22:45–64.

————. 1981. "Ontogenetic Adaptations and Retrogressive Processes in the Development of the Nervous System and Behavior." In *Maturation and Development: Biological and Psychological Perspectives,* eds. K. Connolly and H. Prechtl. Philadelphia: Lippincott.

————. 1982a. "Preformation and Epigenesis in the Origins of the Nervous System and Behavior: Issues, Concepts and Their History." In *Perspectives in Ethology (vol. 5),* eds. P.P.G. Bateson and P.H. Klopfer. New York: Plenum.

————. 1982b. "The Neuroembryological Study of Behavior: Progress, Problems, Perspectives." In *Current Topics in Developmental Biology (vol. 17),* ed. R.K. Hunt. New York: Academic Press.

_____. 1992. "Pathways in the Emergence of Developmental Neuroethology: Antecedents to Current Views of Neurobehavioral Ontogeny." *Journal of Neurobiology* (Special Issue on Developmental Neuroethology) 23:1370–1403.

Oppenheim, R.W. and L. Haverkamp. 1985. "Early Development of Behavior and the Nervous System: An Embryological Perspective." In *Handbook of Behavioral Neurobiology (vol. 8)*, ed. E. Blass, New York: Plenum.

Weiss, P.A. 1955. "Nervous System (Neurogenesis)." In *Analysis of Development,* eds. B. Willier, P. Weiss, and V. Hamburger, Philadelphia: Sanders.

Wilson, E.B. 1900. *The Cell in Development and Heredity.* 2nd rev. ed, New York: Macmillan.

Preface

The past is a refuge for illusions that furnish us with metaphorical justifications for our present course and future existence. One of the illusions grounded in nineteenth-century ideas about evolution that stubbornly persists is the belief that human nature can be best understood as being determined by either hereditary or environmental factors. This dichotomy establishes the limits to scientific inquiry by supplying the universe of meaningful discourse about the range of individual variation possible among the human species. The terms used to characterize the effects of genetic constitution and social contingency on the human personality also assume shifting moral connotations. For example, we no longer hold alcoholics responsible for incorrigibility but treat them for a disease traceable to a genetic defect or developmental disorder.

Myrtle McGraw tirelessly challenged the discourse about development popularized by maturationist theories because it failed to comprehend development in terms of growth. She believed that the processes of growth underpinning development are dynamic and open-ended. Consequently, McGraw argued that the sequence of early development is not fixed but subject to conscious control and redirection. The countless interactions among neural, physiological, behavioral, and other elements form individuals with unique skills. Yet each person possesses the common capacity for judgment and self-transformation. McGraw's insights as an experimentalist in infant development, illustrated by the essays in this book, demonstrate how knowledge about evolution and development can be used as a resource to construct a different future, if we are willing to go beyond the limitations of heredity and environment.

Thomas C. Dalton
Victor W. Bergenn

Acknowledgments

The editors express their appreciation to the following archive collections and libraries for assistance in locating documents and correspondence: The Rockefeller Archive Center, Tarrytown, New York; The Carnegie Corporation, New York City; The Josiah Macy Junior Foundation, New York City; The Center for Dewey Studies and Morris Library, Southern Illinois University, Carbondale, Illinois; Millbank Memorial Library, Teachers College; Butler Library, Columbia University, New York; The History of Science Division, National Library of Medicine, National Institutes of Health, Bethesda, Maryland; The Museum of Natural History, New York; Spencer Research Library, University of Kansas, Lawrence, Kansas; Manuscript Division, The Library of Congress, Washington, D.C.; The Lilly Library, Indiana University, Bloomington, Indiana; and The McGraw Papers, Leonia, New Jersey. Special thanks also to Mitzi D.Wertheim, Lewis P. Rowland, Michael Katz, Ronald W. Oppenheim, J. Lawrence Pool, Theodora Abel, Katherine Heyl, Nancy Harter, Hansl Meith, and Francis Mechner.

T.C.D.
V.W.B.

About the Editors and Contributors

Victor W. Bergenn is executive director of the Council on Educational Psychology (founded by Myrtle McGraw) in Leonia, New Jersey. Bergenn, a former McGraw colleague and lecturer at Briarcliff College, works on a variety of educational assessment projects for state and local agencies, has published in the field of educational testing, and has recently developed clinical measures to diagnose differences in learning styles and readiness. Bergenn is co-chairing, with Dalton, an invited symposium on Myrtle McGraw and the Maturation Controversy for the 1995 meeting of the American Psychological Association.

Thomas C. Dalton is senior research associate and lecturer in politics and society at California Polytechnic State University, San Luis Obispo, and is currently on leave with the Office of the Provost, Arizona State University. Dalton has published works on the history of developmental psychology, Cuban social policies, and has a book in progress on John Dewey's life and work. Dalton is presenting an address to the 1995 meeting of the American Psychological Association on Dewey's collaboration with Myrtle McGraw in the 1930s. Dalton is also a consultant to the producers of *Scientific American Frontiers,* who are preparing a documentary on McGraw's work for the Public Broadcasting System.

Paul Dennis is professor of psychology at Elizabethtown College, Elizabethtown, Pennsylvania. Dennis has published several articles on the history of psychology and is currently conducting research on the history of psychology on radio.

Donald A. Dewsbury is professor of psychology at the University of Florida, Gainesville, Florida. Dewsbury is a comparative psychologist and experimentalist who has published extensively on mammalian behavior and documented the contributions of pioneers in the field of ethology.

Ronald W. Oppenheim is professor in the Department of Neurobiology and Anatomy, Programs in Neuroscience and Cell Biology, Bowman Gray School of Medicine, Wake Forest University, Winston-Salem, North Carolina. Oppenheim is presently conducting research on nerve growth processes and has published extensively on the history of experimental embryology and neuroethology.

Gerard Piel invented *Scientific American*, publishing his first issue in May 1948, and was its first president and publisher until 1984. He is author of several books on issues of "science and society," with the latest, *Only One World*, published in 1992.

Bert C.L. Touwen is a professor of the Faculty of Medicine in the Department of Developmental Neurology, University of Groningen, The Netherlands. Touwen is currently involved in behavioral studies of normal and brain-damaged infants and has published extensively about developmental processes during infancy and early childhood.

Beyond Heredity and Environment

Reconsidering McGraw's Contribution to Developmental Psychology

THOMAS C. DALTON AND VICTOR W. BERGENN

About the Essays and Contributors

Myrtle McGraw was a psychologist who pioneered in the study of infant development in the 1930s and 1940s. McGraw (1935) is best known for her co-twin study of Johnny and Jimmy Woods published in the mid-1930s. Although McGraw's contributions to the field of child development have been readily acknowledged and documented (see for example, Crowell 1967 and Fowler 1983), controversy persists among psychologists as to how to interpret her ideas about significant factors influencing learning processes. Leading textbooks in psychology (see for example, Hetherington and Parke 1993) continue to categorize McGraw as a maturationist, arguing that her findings, like Gesell's, demonstrated that learning readiness and aptitude are governed largely by genetically controlled developmental processes and that early stimulation does not appreciably enhance ultimate performance. The editors critically examine how preconceptions about McGraw's and Gesell's research, including the failure by researchers to fully explore the relationship between brain and behavior, have perpetuated divisive arguments about nature and nurture. McGraw's pioneering research showed how this dichotomy of development could be overcome if developmental biologists and psychologists adopted integrated methods of inquiry.

This volume is intended to address this and related controversies surrounding McGraw's research by bringing together sixteen essays that represent McGraw's most cogent and lucid statement of her perspective. Many of McGraw's earlier articles are not readily accessible to researchers because they were published in diverse professional journals, magazines, or volumes

1

with limited circulation. Therefore, the editors have selected some of her most important early work for publication in a new collection, including six previously unpublished essays and lectures that help clarify her basic concepts and principles of development. These essays demonstrate that McGraw conceived of development and learning in terms of the continuous interaction between neural and behavioral growth processes that could not be reduced to any single specifiable hereditary or environmental factors. The editors contend that the problems of development McGraw incisively illuminates in these essays anticipated the conundrum posed by the insoluble debate over nature and nurture by farsightedly pioneering new methods to comprehend the complex but integrated processes of human development (See Bergenn, Dalton, and Lipsitt 1992 for a detailed examination of McGraw's ideas and research).

In addition, three other contributors provide special introductions, putting in context the themes addressed in the essays found in Parts 1 through 4. Gerard Piel, founding publisher and editor of *Scientific American,* a close friend of McGraw, covered her research as a science reporter. He discusses McGraw's struggle to establish new methods of research in infant development during the 1930s and describes her impact on his career. Paul Dennis, a psychologist and historian, who has recently analyzed news coverage of McGraw's co-twin studies in the 1930s, suggests why McGraw was misunderstood and examines her subsequent attempts to correct misinterpretations of her work. Thomas Dalton, a political scientist and historian of the social sciences, explores the significance of Coghill's neuroembryological concepts in McGraw and Gesell's experimental work, and identifies important but overlooked differences in their conceptions of child development. Finally, Donald Dewsbury, a comparative psychologist, assesses McGraw's contribution to the field of psychobiology and examines the merits of the arguments she makes for the interdisciplinary study of developmental processes.

The editors provide an introductory essay that reconstructs the context in which McGraw conducted her studies at Columbia University. We demonstrate that her work was the product of a little known interdisciplinary collaboration with several scientists that included the philosopher John Dewey, Frederick Tilney, a neurologist, and George Coghill, a neuroanatomist. Dewey spent a lifetime arguing against dualisms such as that embodied in the heredity-environment dichotomy. Dewey (1935b) believed that McGraw's work promised to "revolutionize work in the field of child study," and that the biological principles she elucidated "prepare the way for deliberate inquiry," and "foreshadow its pattern"—an argument that Dewey (1986:30) ultimately advanced in *Logic: The Theory of Inquiry* in 1938 (See Dalton and Bergenn 1994 for a more detailed account of their collaboration). Their collaboration brings to light new evidence that her research not only involved the use of novel methods to study growth but enabled Dewey to examine the origin and role of judgment in inquiry.

McGraw's methods also embodied Coghill's controversial but generally misunderstood ideas about neuroembryological growth. The editors draw on documents from the Gesell and McGraw archives, correspondence between Dewey and McGraw, and previously restricted interviews to identify the role that Dewey, Coghill, Lawrence Frank, a foundation official and Dewey colleague, and other associates played in McGraw's infant studies. We also describe how a professional rivalry between McGraw and Gesell (and their struggle to attract Coghill's attention and support) contributed to key differences in their methodology and findings that have generally escaped the attention of serious researchers.

The maturation controversy has never really subsided since McGraw and Gesell's pioneering studies. We critically examine the contours of an ongoing debate between experimental psychologists Philip Zelazo and Esther Thelen, because their disagreements about the role of the brain in behavior effectively illustrate why McGraw and Dewey believed that the brain and behavior function integrally throughout development. In addition, we look at the recent work by paleobiologists Michael McKinney and Kenneth McNamara (1991) who propose a highly suggestive theory about growth processes that demonstrates the pertinence of McGraw's research in current endeavors to understand human evolution. Research mounted by developmental neurologists Heinz Prechtl (1989), Bert Touwen, and their colleagues, in our estimation, significantly extends lines of inquiry closely related to McGraw's research. We also suggest how neuroscientist Gerald Edelman's (1989) seminal research on the brain and consciousness gives new meaning to McGraw's little known findings about memory and recall in her follow-up study of Johnny and Jimmy. Finally, we briefly describe McGraw's research during her years at Briarcliff College from the 1950s to early 1970s, to illustrate why she believed communication to be indispensable to infant development and learning processes.

The editors present McGraw's essays in four parts. Part 1 includes an unpublished lecture McGraw presented to Ronald Oppenheim's colleagues and students in 1973 that provides humorous anecdotal observations about her early career, candid observations about Freudianism and Behaviorism, including the work of her rival, Arnold Gesell, and thoughts about the future of developmental studies. We have also included an unusual essay by McGraw (1985a) recounting several "blunders" she confessed to having committed that may have contributed to the confusion and controversy surrounding her work. This candid article provides an excellent point of departure as subsequent essays focus on different dimensions of the so-called blunders, such as her failure to clear up ambiguity surrounding her notion of "critical period," to mount a more systematic critique of the "maturation" thesis in the decades following her research, and to advance a systematic theory of growth and development.

Part 2 includes three essays published several years after her co-twin study that effectively address misconceptions about her work and clarify key concepts. McGraw's follow-up study of Johnny and Jimmy in 1939 (see Chapter 3) demonstrates that Johnny retained the advantages of his early stimulation. Her study also showed that specific attitudes Johnny exhibited with the attainment of neuromuscular integration seemed to also account for Jimmy's ability to eventually complete tasks he previously found difficult or impossible to accomplish. McGraw's 1940 essay on "Signals of Growth" is included because of its clear illustration of critical periods in learning and the constructive role of temporary regressions in early development. In addition, the editors include two unpublished papers, "Experimental Twins," later revised and published in the *New York Times* in 1942, and "Elkind's Blunders," an essay McGraw sought unsuccessfully to publish in 1986. In these essays, McGraw rebuts the claim that she trained or conditioned the twins to perform selected tasks or that her experiments altered the twins personalities. Finally, McGraw's lecture on "Infant Motor Development" presented in Moscow in 1958, clarifies her findings about the effects of early exercise on subsequent development.

In Part 3, the editors include an early article on reflexes and the first and last chapters of *Growth* that illustrate how McGraw interpreted and applied Coghill's neuroembryological concepts in her infant studies. Understanding Coghill's ideas is important as his work continues to be of interest to neuroscientists and psychologists involved in the study of early development. The chapters from *Growth* are significant for two additional reasons. First, McGraw advanced a philosophy of science (see below and Chapter 9) that foreshadowed Dewey's argument in *Logic*, by suggesting an analogy between theoretical development in psychology and an infant's attempt to attain erect locomotion.

Second, McGraw employed a weaving analogy to describe the relation between infant growth and development (see Introduction to Part 3 and Chapter 10) four years *before* Gesell (1939) proposed his conception of development as a spiral process of "reciprocal interweaving." The chapter on "Individual Development" from her book, *The Neuromuscular Maturation of the Human Infant* is also included here because of its central bearing on the attempt by McGraw and her associates to devise growth constants that would enable quantitative analysis of transformational processes occurring during human development. Finally, having applied Coghill's methods in her studies of learning processes, McGraw demonstrated, in a 1943 lecture, "Let Babies Be Our Teachers," delivered to the New York Academy of Medicine, the additional insights she gained from Coghill for enhancing the prospects of more effectively organized social systems.

Finally, three essays in Part 4 illustrate McGraw's impressive range and sophisticated level of understanding about the neurobiology of development. These essays clearly demonstrate the pervasive influence of Dewey's conception of inquiry as in "Basic Concepts and Procedures," indicate McGraw's farsighted grasp of the experimental literature and issues that remain timely for developmental research today, as illustrated in "Maturation of Behavior," and offer insightful guidelines for the interdisciplinary study of developmental processes, as shown in "Challenges to Students of Infant Development." An unpublished essay McGraw wrote in 1969 is included (see Chapter 15) because of her cogent critique of Freudian and Behaviorist theories and succinct statement of the importance of communication in her theory of child development.

In his foreword, Ronald Oppenheim, a developmental neurobiologist, friend, and colleague of McGraw, assesses the "nascent" and important contribution McGraw has made to contemporary research by adapting the methods of developmental biologists to the study of infant behavior. In addition, Professor Bert C. L. Touwen, a developmental neurologist, describes in his afterword specific aspects of McGraw's research that have been useful in his own studies and to scientists studying the neurological, behavioral, and clinical dimensions of infant development.

Putting McGraw's Research in Context

None has better captured the timelessness of Myrtle Byram McGraw's life and work than her daughter, Mitzi Wertheim, who said at a memorial held at Columbia University in 1988, shortly after her mother's death, that her "mother was born at the end of the nineteenth century, she lived in the twentieth century, and she thought in the twenty-first century" (Lipsitt 1990:977). Indeed, McGraw was born into a world and educated by teachers enthralled by the scientific implications but troubled by the moral dilemmas posed by nineteenth century science, dominated by Herbert Spencer and Charles Darwin's theories of evolution. McGraw grew up in an America whose parents were preoccupied with genetic influences on childhood, seeking unequivocal guidance as to whether nature or nurture was more important in shaping temperament, learning, and behavior. But McGraw's efforts to demonstrate experimentally the reciprocal relationship between brain and behavior—between maturation and learning—forced her to conceive of methods beyond the technological grasp and theoretical comprehension (but now within reach) of most of her colleagues and successors in this century. That is why McGraw's legacy has yet to be fulfilled.

McGraw's rather inconspicuous life as an Alabama schoolgirl took a significant, if not fateful turn when she wrote John Dewey as a teenager from 1914 to 1918. The letters they exchanged subsequently disappeared, but

Dewey's correspondence with McGraw during the 1930s and early 1940s documents their close personal ties and remarkable collaboration in her infant studies. Dewey reciprocated for McGraw's evident awe and admiration by acting, as she recalled, as her "intellectual godfather" (McGraw 1990:934). Dewey remained in contact with McGraw while she attended Ohio Wesleyan, spoke at the school, and influenced her decision to study psychology at Columbia University, where Dewey was professor of philosophy. McGraw took several of Dewey's courses, while supplementing her study in psychology with classes in neuroanatomy before obtaining her Ph.D. in 1931. Dewey was one of the members of her dissertation committee. McGraw (1967:6) characterized her association with Dewey as "a very devoted sort of father-daughter relationship" in which she was considered a part of the family. McGraw recalled that Dewey suggested many of the ideas she tested in her research and said that "I discussed everything I did with him" (McGraw 1972a:32).

Before the turn of the century, Dewey posed questions about evolution, biological growth and development, and the relationship between the brain and intelligent behavior that went well beyond available knowledge and methods. For example, Dewey (1976b:15) speculated that the competing perspectives of morphololgy and physiology could be reconciled in experimental studies because he believed that biological structures and functions evolved through "a common chemico-physical process." Dewey (1976b:310) also boldly asserted in *Studies in Logical Theory* in 1903 that "psychology as the natural history of the various attitudes and structures through which experience passes" was "indispensable to logical evaluation" because it provided the mode in which logic could be reconstructed.

However, the philosopher Charles Peirce criticized Dewey for failing to show how physiology or psychology could possibly elucidate the evolution of function or how comparative anatomy could be applied to the "anatomy of thought" (Burks 1958:181). Peirce also chided Dewey by saying: "There is no anatomy of possibilities because no one can say in advance how pure possibilities vary or diverge from one another" (Burks, 1958:181). Interestingly, McGraw anticipated this problem by devising a means to reveal the variety of behavioral forms *hidden* in the developmental process.

Dewey had previous occasions to observe and analyze infant and child developmental processes. At his Laboratory School at the University of Chicago (1896–1904), Dewey showed a surprising acuity in identifying sequential stages of sensori-motor development. He hypothesized that attention and awareness come into play when infants attempted to attain balance in a sitting position, preparatory to prone locomotion, and observed that the capacity for recognition and comparison seemed to accompany crawling and creeping (Dewey 1976a:184; 187–188). Dewey concluded from his research that an experimental program could be devised "to discover some single continuous function undergoing development in order to bring scientific relevancy and order into various facts of child psychology, and in order to give them practical or pedagogical usefulness" (1976a:191). McGraw ultimately devised such a program, with Dewey's help, years later in her studies of infant locomotion.

Dewey was also on the board of directors (1916–1920) of the Bureau of Educational Experiments begun by Lucy Sprague Mitchell and her husband, Wesley, in 1916. The Bureau sponsored an exploratory (but not experimentally controlled) attempt over a four year period to measure the effects of stimulation on rate of growth according to age, intelligence, and behavior (Antler 1982). Although suggestive, the results were inconclusive (Johnson 1925). Bureau staff lacked the skills necessary to obtain a detailed record of the sequence of development. They also found it increasingly difficult to compete for foundation support with Arnold Gesell, whose motion picture studies of the genesis of infant behavior were already underway at Yale (Mitchell, 1925).

McGraw's studies provided Dewey the opportunity to ground his psychological premises about judgment in inquiry in the latest scientific evidence regarding the neurogenesis of behavior. Dewey had largely failed to convince his philosophical colleagues that intelligence evolved to meet the demands for coordinated action and that erect locomotion gave birth to inquiry by enabling early humans to anticipate and control the consequences of their actions (see generally, Nagel 1977; 1986 and Russell 1977). Consequently, Dewey wanted to identify the circumstances that contributed to the need for judgment, to isolate the specific traits involved, and to show how they helped form the pattern of inquiry. This knowledge would enable Dewey to explain how propositions enter existentially into inquiry through methods, bringing events under control, while increasing the ability to meet new contingencies (see Dalton and Bergenn, 1994).

Dewey also struggled unsuccessfully to demonstrate why ideas do not depend on the association of stimulus and response but originate indirectly from feelings, rooted in organic processes of suggestion that prolong judgment and control over the direction of inquiry. Dewey believed that inquiry was not limited to thinking or discourse, but involved integrated psychobiological processes. That is why McGraw focused on how infants' attitudes reveal (and affect) the level of awareness, deliberateness, interest, and resistance involved in getting a problem under control. Thus Dewey argued that effective inquiry depended not on the accumulation of confirming evidence through redundant behavior or operations. Instead, Dewey contended that inquiry involved the continuous modification of our attitudes and methods, and the reintegration of knowledge and experience, expanding the scope of human judgment and understanding.

Dewey (1930) privately acknowledged an intellectual debt to Charles M. Child (1921), a neurophysiologist, and C. Judson Herrick, a neurobiologist, both close personal friends from the University of Chicago, for focusing his ideas about the role of the brain in judgment. Child argued that metabolic gradients formed in embryogenesis accounted for neural and behavioral patterns formed during subsequent stages of development. This led Dewey to speculate that the processes through which ectoderm and mesoderm interacted furnished the best analogy for understanding the structure of inquiry. Herrick (1910a; 1910b; 1913; 1962) found neurological evidence supporting Dewey's contentions that neural mechanisms of inhibition and anticipation made possible intelligent behavior. Herrick's search for evidence to support Dewey's conception of coordinated intelligence led Herrick (1949) to formulate a highly original psychobiological theory of human evolution. (see Kingsland 1990; 1993). Herrick observed McGraw's research at Columbia on at least one occasion and also provided consultation on her early publications (McGraw 1932).

Dewey became familiar with Coghill's work through Herrick, who collaborated in some of Coghill's early studies. Coghill (1933a) found Dewey's criticisms of stimulus-response theory and conditioning very persuasive and explicitly acknowledged his intellectual debt to Dewey on several occasions (Herrick 1949). Coghill argued that early neural growth anticipated functional development and that complex behavior presupposed neuromuscular integration. Coghill became actively involved in McGraw's studies to see if his theory applied to human development. Herrick and Coghill each supported and assisted Dewey (1986:40) in his attempt to find evidence through McGraw's studies to support his contention in *Logic* that human intelligence evolved from a long series of ontogenetic adaptations, involving the continuous reintegration of the structural and functional traits of human existence.

Dewey was actively involved in McGraw's Normal Child Development Study (NCDS) throughout the 1930s and early 1940s. Dewey served on an advisory committee chaired by Dr. Frederick Tilney, a neurologist, that included Dewey colleagues, psychologists Robert Woodworth and Edward L. Thorndike and several other physicians and scientists from Columbia University. Coghill served as an ad hoc member while employed at the Wistar Institute in Philadelphia, as did noted behaviorist John B. Watson, whose involvement was limited to occasional attendance at advisory meetings. Coghill provided timely methodological advice and consultation in the early stages of McGraw's experimental studies (see Introduction, Part 3). The Advisory Council of the NCDS met regularly throughout the 1930s to discuss presentations by Advisory Council members such as Dewey and Tilney, interim progress reports, and briefings or lectures by staff. Dewey and McGraw also participated in informal discussions with Advisory Committee members on Thursday evenings which were occasionally attended by Margaret Mead (McGraw 1972a:41; McGraw 1967:17–18). A former McGraw associate Katherine Heyl (1989) who worked on the NCDS from 1933–1936, recalled that Dewey had an office at Babies Hospital and visited on a daily basis. Heyl was under the impression that Dewey had originated the project because McGraw consulted with him about it frequently. Indeed, McGraw recalled years later that: "Every now and then I wake up to the fact that some idea that I think was my own, if I happen to pick up something he wrote long ago, he was saying it. My connection with him was just learning by living and by talking" (McGraw 1972a:32).

Tilney appointed McGraw associate director of the NCDS in 1930. Tilney was the first professor to teach courses in neuroembryology when the field was in its infancy and directed research at the Neurological Institute of New York before becoming its head in 1935 (Elsberg 1944; Pool 1975). Tilney (1968) had an absorbing interest in the relationship between brain and behavior. He extrapolated brilliantly although speculatively from neuroanatomical and museum specimens about the evolution of the brain. Tilney (1923) suggested, for example, how cerebral structures were likely to have evolved in response to changing functional needs and demonstrated a sophisticated understanding of psychological studies and behaviorism (Tilney and Casamajor 1924). He used innovative techniques to determine if the sequence of neural maturation was correlated with the development of behavior. Tilney also outlined the role that the NCDS would play by noting that "the critical phases in the development of child behavior will be indicated and structural studies of the brain made to accord with them chronologically" (Tilney and Kubie 1931:239).

McGraw (1935:10) readily adapted Tilney and Coghill's methods in her infant studies, contending that "it is the experimental embryologists, not psychologists who deserve credit for formulating the most adequate theory of behavior development." McGraw (1979) acknowledged Coghill's extensive influence by saying that: "Coghill visited my laboratory many, many, many times—sometimes with Tilney, sometimes not. We talked and exchanged ideas. It was he, John Dewey, and the babies that got me thinking of process, not end result, or achievement." Indeed, McGraw found evidence that infant behavior develops through processes of differentiation and integration similar to those Coghill observed in the fetal development of salamanders. McGraw (1979) also stressed that collaboration among them was essential to the success of the project by saying: "Had he [Tilney] lived longer, Tilney, Coghill, and Dewey and I (let me say the babies) might have arrived at a synthesis of the meaning of structure and function."

Aside from Dewey, Coghill, and Tilney, no other figure looms larger in McGraw's seminal research on child growth and development than Lawrence K. Frank. He came to the field of child study in the early 1920s as a former Dewey student, and soon occupied strategic positions with the Rockefeller Foundation and the Josiah Macy Jr. Foundations from the mid-twenties to early forties. Frank's gift was his unrivaled ability to identify bright, talented people and orchestrate their work into a harmonious symphony of ideas saturated with insight (Frank 1962). Frank's organization of an international and interdisciplinary network of academically based researchers demonstrated a foresight unparalleled today (Cravens 1993). Frank sought to create through this network of academic institutes (at Columbia, Iowa, Berkeley, Minnesota, Yale and other universities) an alliance between parent groups and researchers that would disseminate knowledge widely about child development, arousing public support for more innovative approaches to learning and child rearing (Dalton 1994).

Frank possessed a sophisticated theoretical grasp of developmental biology and psychology and contributed important articles and books to the field of child study (see for example, Frank 1935a; 1935b; 1949). Frank considered the analysis of growth to constitute an essential focus of developmental studies and defended the need for McGraw's studies despite the growing insistence that research of this kind yield diagnostic tools for clinicians. Frank urged support of McGraw, even though the Rockefeller Foundation was already funding Gesell's studies at Yale, because he contended that McGraw was not establishing age norms but trying to "delineate the sequence through which the child passes and to discover how far that sequence is modifiable by training" (Frank 1933).

Shortly before leaving the Rockefeller Foundation in 1936 to join the Josiah Macy Jr. Foundation, Frank helped Tilney and Willard Rappleye, Dean of the College of Physicians and Surgeons at Columbia University, to obtain a large multi-year grant from the General Education Board, supporting the continuation of the NCDS and related studies. This was an ambitious proposal to analyze all major structural and functional elements of growth including anatomical, physiological, and psychological dimensions (Rappleye 1935). McGraw outlined her own work program in some detail, proposing to gather and correlate quantitative data about neuromuscular mechanisms and other developmental processes, involving the brain, respiratory system, and metabolism. She also proposed studies of infant vocalization, gesture, and concept formation. With Frank and Dewey's assistance, McGraw put together an interdisciplinary group of 10 researchers and 8 technicians that included: a neurophysiologist, a physiologist, a biochemist, 2 pediatricians, 3 psychologists, and 2 nurses. The complete scope and sophisticated nature of this research documented in over 50 journal articles has yet to be fully appreciated!

Dewey was a founding board member (1930–1944) of the Josiah Macy Jr. Foundation that supports medical research (Rappleye 1955:ix)—a fact unbeknownst to Dewey's closest colleagues and one that has eluded researchers. Dewey, and Frank, who served as vice president from 1936–1941 under Ludwig Kast (1930–1941), had an unusual opportunity to influence the scope of McGraw's studies as well as to participate in the selection of many other collateral research projects funded during the same period (Kast 1936:34–37; Rappleye 1955:ix). Dewey (1988a) shared Kast's (1936:11) interest in reforming medical practice by getting physicians to adopt integrated psychobiological approaches to preventative health care in medical schools.

Dewey and Kast assisted Tilney in seeing that the NCDS was effectively coordinated with other research projects recommended by the Carnegie Institute of Washington, so that human development could be studied from the broadest possible perspective (Merriam 1933; Tilney 1933; 1934). In addition, Dewey's daughter Evelyn was commissioned by the Macy foundation in 1933 to prepare a report in which she urged "a more systematic review of existing knowledge, especially with respect to fetal and infant development," noting that McGraw's studies were organized to increase knowledge of the relation between brain and behavior (E. Dewey 1935:354–356).

GESELL, THE MATURATION CONTROVERSY, AND AFTERMATH

As Paul Dennis indicates in his intriguing introduction to Part 2, no other psychological research was covered more continuously by the press and popular magazines than McGraw's experimental studies of Johnny and Jimmy Woods. The debate over whether behavior was influenced more by heredity or environment, fueled by the rivalry between John Watson and Arnold Gesell, aroused a palpable sense of expectancy among the public that McGraw's research would decisively resolve the issue (Dennis 1989). The personification of the debate in two irresistible youngsters fed an insatiable curiosity among a public anxious to know the limits of training for learning and personality development. Given this atmosphere, it is not surprising that McGraw's most significant discoveries were overshadowed, because they could not be fit easily into the dichotomy of nature or nurture in development.

Although Gesell and McGraw sought to transcend in their research the narrow confines of the nature-nurture debate, they were engulfed by it, contributing to their rivalry and diffidence. As a student, McGraw respected and admired Gesell's (1934) work. She came away from her first visit at Gesell's clinic in 1926 "with a renewed professional enthusiasm," reminding Gesell that, "It is after all this exchange of ideas that makes the wheels go round" (McGraw 1926). McGraw (1927) followed up her visit the next year by unsuccessfully seeking employment on Gesell's staff. However, once McGraw assumed her position on the NCDS, she found it increasingly difficult to maintain cordial relations with Gesell. Gesell was justifiably put off by Tilney sending his subordinate, McGraw, to review his experimental procedures (see McGraw 1972a), but acted unfairly by never citing any of McGraw's research—a slight that disappointed and annoyed McGraw—unwittingly contributing to the continuing confusion regarding their respective views (McGraw 1980:15–16). Moreover, McGraw resented Gesell's medical degree, which opened doors and commanded respect denied her as a female psychologist.

John Dewey seems to have shared in this rivalry by the tone of his references to Gesell. For example, Dewey (1934a:4) reported to McGraw a "disgusting and amusing" incident with Gesell while on board a ship headed for an educational conference to Capetown, South Africa. Dewey, who considered modesty a virtue, typically shunned posturing and affectation associated with celebrity. Apparently Gesell and Dewey were recognized by a maitre d' who asked Gesell if Dewey "was as celebrated as Gesell" and subsequently told Dewey that Gesell replied by saying: "Oh, much more so" (Dewey 1934a:4).

Having established his credentials as a leading educational reformer long before Gesell entered the field, Dewey believed that McGraw's research strongly supported his previously untested theories about the role of judgment in learning and behavior development. Dewey argued against the idea that humans possessed unalterable, permanent traits and that human development unfolded in a predictable and invariable sequence. Instead, he believed that the form and function of human behavior depended entirely on contingencies of order, and that the attitudes we adopted could decisively affect conceivable and attainable forms of human existence. Dewey (1935a:viii) praised McGraw's work in his introduction to *Growth*, declaring that she had tentatively established general principles of child development. Dewey also considered her achievement comparable to that of Michael Faraday, a physicist and nineteenth century pioneer in field theory (New York Times 1935:19) because she demonstrated that growth processes advance like interpenetrating electromagnetic waves, redistributing energy through overlapping neuromuscular phases of development that enable behavior to assume new forms.

Since this exhilarating period of discovery, there has been a discernible positive trend towards reformulating the discourse about child development in interactionist terms (see for example, Gottlieb 1991). Nevertheless, some experimentalists in infant development and learning have shown little inclination to put the nature versus nurture dichotomy to rest, but instead, have resurrected it in new guises. Esther Thelen, an experimentalist in infant motor development, has become a prominent opponent of behavioral genetics which she criticizes as deterministic. She is also an outspoken critic of neural behavioralists, whom she contends, believe that the human brain is "hard-wired" with a pre-existing set of instructions that govern development. Thelen (1990:22; 34–35) defines her own approach as a "dynamic systems theory" which she claims avoids reducing infant behavior to genetic or environmental factors by treating neuromuscular processes as if they were "self-organizing systems." Thelen's interpretations of the work of McGraw and Gesell are sometimes contradictory. Therefore, a close scrutiny of her portraits of these pioneers (especially McGraw) is in order not only to set the record straight, but to suggest why the *integrated* study of brain and behavior offers the best prospect for future breakthroughs in knowledge about child growth and development.

Thelen has not been altogether consistent in how she contrasts and appraises Gesell and McGraw's contributions. In one of Thelen's accounts, McGraw is compared unfavorably to Gesell for having adopted, according to Thelen (1987:10;12) "a more prescriptive neural-maturationist model" that is akin to the reductionist tradition in biology. Gesell's ideas, in contrast, "are worthy of serious reconsideration," according to Thelen (1987:10), because his emphasis on self-regulating functions and avoidance of "the apparent linearity of McGraw's conceptualization" is closer to a dynamical systems view (Thelen 1987:14). Yet in another separate evaluation of Gesell's work, Thelen and Adolf (1992:376) criticized Gesell for explaining all behavior through postural mechanisms and "doggedly assigning the intricacies of development to a single cause," which is genetics. In this instance, she concludes that McGraw's "documentation of early motor development was equal and in may ways superior to Gesell's in technical and descriptive elegance" (Thelen and Adolf 1992:376).

Thelen correctly observes that McGraw studied movement because she believed that the form exhibited in behavior appearing in different stages of locomotion would reveal the underlying level of neural maturation. However, Thelen (1987:6) interprets this incorrectly to mean that McGraw supposed that all behavior was determined solely by processes of neural maturation, governed by the cerebral cortex, and that consequently, Thelen contends, McGraw believed that "function emerged from structure and not the reverse." McGraw argued instead (see Introduction to Part 3) that each phase of locomotion exhibited a different form or pattern largely because of variations in the rate of growth and timing or interactions among neuroanatomical structures, physiological, and behavioral functions. She demonstrated that neural structures sometimes advance more rapidly, impeding processes of functional consolidation, while at other times, new functional capabilities outstrip the possibility of effective coordination, contributing to regression and the diminution of conscious control. This hardly squares with Thelen's view that McGraw collapsed function into structure or that she advanced a unilinear theory of development.

Thelen (1987:13) is anxious to paint McGraw and other proponents of the so-called "top-down" explanations such as Zelazo into a corner because she believes that "neural maturation alone is an impoverished basis for a developmental theory," suggesting that: "In many ways, Coghill's correlative paradigm remains unfulfilled" (Thelen 1987:13). Upon closer inspection, however, the research program that Thelen is embarked on is dedicated to the elimination of any references to developmental brain states, consciousness, or intentionality in her ambitious but incomplete line of inquiry. Thelen contends that her dynamic systems theory offers a more "parsimonious" account of developmental behavior because no *post hoc* references to pre-cognitive states, neural remodeling or programming, or fetal behavior is needed.

Thelen claims that different motor repertoires characteristic of behavioral development can be explained solely by changes in the distribution of energy and relationship of forces characteristic of all biomechanical systems. However, Thelen has not yet formulated mathematical models that provide a precise description of how energy is converted through muscular processes alone to allow functional behavior to assume new forms. McGraw and her associates demonstrated with some limited success correlations between neural growth processes and the sequence of transformation of motor patterns in early development. They also argued that a complete explanation of developmental processes must demonstrate how brain and behavior interact both quantitatively (i.e., with respect to biological processes) and qualitatively (i.e., through consciously deliberate acts) to redistribute energy to support new and more complex behaviors.

The Significance of Erect Locomotion in Human Development and Inquiry

Zelazo (1983) and Thelen (1983; 1987) are engaged in an ongoing controversy about the significance of early motor patterns in human evolution and cognition, which involves important issues with deep historical roots. Zelazo (1983) contends that infants' early rhythmic stepping movements reflect a rudimentary capacity for deliberation and that the aided exercise of these behaviors stimulates memory and recognition—capacities conducive to the attainment of erect locomotion. Zelazo (1983:133) frames the implications of this discovery very succinctly by saying: "Independent erect locomotion is unique to man, and if ontogeny approximates the recapitulation of phylogeny, as is often stated, then perhaps a fragment of man's uniqueness can be uncovered from more intensive study of the infant's assumption of unaided walking. To date, we have glanced over this event too casually." Thelen (1983) rejects Zelazo's argument, contending that attainment of erect locomotion does not imply volition nor require cognition. Thelen argues that kicking and stepping are stereotypical movements made possible solely by the removal of gravitational constraints and such movements soon disappear. Consequently, she argues that it is more likely that bipedalism evolved in conjunction with anatomical and subcortical adaptations precipitated by environmental changes, long before the expansion of the neocortex (Thelen 1983:153–154).

McGraw had the occasion to correspond with Philip Zelazo after he and his associates (Zelazo et al. 1972b) published the findings of his studies that demonstrated that stimulated reflex stepping facilitated walking in infants. Zelazo (1972a) asked McGraw whether Johnny (the twin that received special stimulation) had received any stimulation of the stepping reflex during the first two weeks of life and asked whether the general exercises he was provided made any difference as to when he began to walk. McGraw's response reveals her genius for improvisation:

> During the reflex period we merely held him under the arms to elicit the reflex stepping. As he approached what I call the transition from reflex to deliberate stepping we put him in a little harness with straps extending to a rod above the table. The straps were attached to ball-bearing rollers so that whenever they were kicked they moved forward. In other words, the action was his. He certainly did a lot more stepping than his twin brother or the other babies. Most of them simply stood on their toes or feet with the shoulders forward, and you can't make them step for love or money during this phase of the activity (McGraw 1972b).

As for the effects of stimulation, McGraw was candid in acknowledging that:

> I completely agree with you that I can never know whether the exercise advanced the age of Johnny's independent walking. As an old scientist, I can only say that I never had direct proof of it. Between you and me I always suspected that it did. I do know that exercise of motor activities, once the motor area of the cortex comes into play, makes an enormous difference in the child's general psychomotor performance. I followed the boys in the laboratory until they were nearly 10 years old. ... I have some non-experimental films taken even later. [McGraw is referring to a film she took of Johnny and Jimmy when they were 22 years old in 1952.] I will categorically say that Johnny ended up a much better coordinated adult in motor performance, and I attribute that superiority to the type of exercise he received, when the motor area of the cortex was maturing during the first two years of life (McGraw 1972b).

Zelazo et al. (1993:690) have since demonstrated that the "effect of practice on neuromotor development is specific to the pattern trained. Stepping practice, alone or in combination with training of sitting, led to increased stepping." But this facilitative effect did not carry-over to the neuromotor

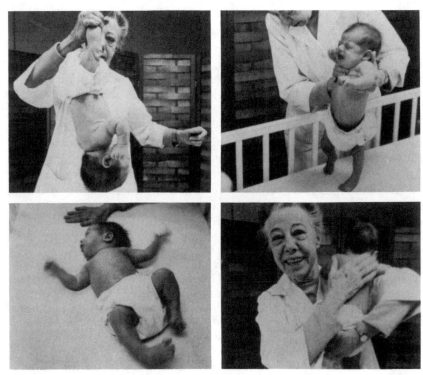

McGraw demonstrates (*top left*) that at 4 weeks an infant, unlike other animal young, doesn't know it is upside down and does nothing to right itself. An infant is capable of reflex stepping (*top right*), a capability that disappears at 3 months. However, unlike an adult, an infant reacts throughout body, arms, and legs (*bottom left*) when startled by pounding on the bed. After testing, McGraw comforts the infant. (Photo courtesy of Francis Mechner.)

pattern of sitting. Nevertheless, Zelazo and his associates contend that practice of prelocomotor movements, such as stepping, prolongs their influence, making them accessible to control and incorporation later into more complex behaviors. Dalton indicates in an introductory essay to Part 3 that McGraw found similar experimental evidence that Johnny successfully transferred techniques, such as those introduced to attain balance on a stool, to dismount it more effectively than controls.

Dewey, Tilney, and McGraw raised fundamental questions about the ontogeny of locomotion and its importance to judgment and inquiry that put the issues debated by Zelazo and Thelen in sharper focus. Dewey speculated in two seminal books in the 1920s that erect locomotion marked the birth of human inquiry. For example, in *Human Nature and Conduct*, 1922, Dewey makes the following remarkable observation:

> The habit of walking is expressed in what a man sees when he keeps still, even in his dreams. The recognition of distances and directions of things from his place at rest is the obvious proof of this statement. The habit of locomotion is latent in the sense that it is definitely at the fore. But consideration is not suppression. Locomotion is a potential energy not in any metaphysical sense, but in the sense in which potential energy as well as kinetic has to be taken account of in any scientific description. Everything that a man who has the habit of locomotion does and thinks he does and thinks differently on that account. This fact is recognized in current psychology, but is falsified into association of sensations (Dewey 1988:29).

Dewey (1988b:135) goes on to say that: "Deliberation is an experiment in finding out what the various lines of possible action are really like. It is an experiment in making various combinations of selected elements of habits and impulses, to see what the resultant action would be like if it were entered upon." These statements underscore Dewey's supposition that thought and action are intertwined. McGraw saw infant motor behavior largely in the same terms as attempts to learn by drawing out and elaborating the preliminary physical postures and movements involved in an activity. Dewey (1981:197–209) argued in 1925 in *Experience and Nature*—a book McGraw (1972a) recalls typing for Dewey—that if we were to retrace the steps involved in the evolution of locomotion, it is likely that we would find that the capacity for comparison, foresight, and anticipation entailed in judgment, found full expression in the transition to bipedal locomotion.

Tilney (1968) addressed the question of locomotion from a comparative perspective, suggesting that the liberation of hands from feet, making bipedalism possible, enabled early man to grasp and manipulate objects and reason. Tilney leaves no doubt that the brain contributed to rather than followed this momentous achievement by observing:

> Summarized as briefly as possible, it may be said that what the brain owes to the hand and foot is the frontal lobe. Through all the stages of progress, from the time when the monkeys first began to live in the trees until their successors, through graded intermediate phases, developed the hand and foot of man, this lobe has been the outstanding feature of the brain. ... It is perhaps unwise and also unwarranted to speak of the debt that one organ owes to others, especially when the activities all represent unified processes, but it was the opportunities presented to hand and foot that amplified the brain (Tilney 1968:299).

Tilney's views resonated clearly in McGraw's memory years later. In a lecture at Teachers College in 1985, McGraw recalled being asked by an Australian participant at a conference she attended at Oxford in 1976 why she focused on the study of motor development; she responded that "it was the easiest to observe." She then related in the following anecdote the evolutionary significance of her remark by adding:

> But there is something more to it than that. In the course of acquiring erect locomotion, freeing the forelimbs for manipulation rather than participation in the total body movement, is one significant indication in supporting the evolution of man. The infant does not acquire these distinctions all at once. Indeed he begins to use his arms and hands manipulatively (reaching out and handling objects) before he is freed completely from using the forelimbs for body movement. All too often, we investigators overlook the depth of significance in the most common behavioral development. Interestingly, adults tend to swing their arms in harmony with the movement of the opposite limb if they are not using the arms to hold something. Perhaps this is a non-essential residual of quadrul locomotion in the mammalian species. If the time ever comes when such qualitative changes in normal everyday behavior can be identified with neurons and the circuitry of the brain and nervous system it will be an enormous achievement in neuropsychology (McGraw 1985b:5).

McGraw's observation is important because it explains why she considered crawling and creeping to constitute such significant events in ontogeny. It is precisely at these stages of infant locomotion, according to McGraw (1941a), that an infant attains sufficient balance to coordinate arms and legs in an *integrated* form of behavior. The synchronous movement of opposing arms and legs in crawling furnishes the balance needed to ultimately transcend the limits of prone locomotion. Only after mastering erect locomotion is an infant fully able to move arms and legs independently in order to use them in conjunction or separately to perform different operations involved in inquiry.

THE COMMON PATTERN OF DEVELOPMENT AND INQUIRY

McGraw's (1942) experimental studies furnished evidence that awareness flickers at the earliest stages of infancy, sometimes exhibited by an urge for propulsion or in a suggestive attitude of curiosity or arousal. Importantly, McGraw discovered that infants exhibit attentiveness and deliberation when they first begin to crawl and explore their immediate environment. McGraw (1935:306–308) also observed that growth processes are not straightforward. Developing behavior patterns alternate and overlap, according to McGraw, pulsating or oscillating forward and backward in rhythmic waves. Development proceeds through growth phases, involving exaggerated and inhibited movements, the elimination of excess motion, and the consolidation and integration of complex behaviors. This alternating sequence, in-

volving frequent reversions to more rudimentary behavior, seems paradoxi-
cal because individual variations in the timing, composition, and direction
of movements occur without altering the general pattern of progressive development. Nevertheless, with each backward swing, older traits and emergent capabilities are actually recombined and reintegrated in slightly different ways, according to McGraw, to alter the form or configuration of a total
pattern. Consequently, excessive or idiosyncratic elements of a behavior pattern are eliminated in the course of development at the same time that new
traits and functions are substituted that contribute to greater flexibility and
adaptability (McGraw 1939).

The importance of McGraw's principles of development in Dewey's analysis of judgment in *Logic* cannot be overstated. The distinctive pattern of development McGraw discovered furnished analogies that Dewey used to further elaborate the structure of judgment and function of inference in inquiry.
These principles led Dewey (1986:197–199) to argue that reasoning did not
conform to the absolute and nonreversable rules of identity and contradiction but followed an indirect course, involving the formulation of functionally specific responses to contingencies in human development. The alternating phases through which development progresses, forming the existential
matrix of inquiry, Dewey believed, demonstrated that judgment emerges
through overlapping stages of comparison through contrast, reorientation,
and redirection that increasingly delineate the scope of effective action. Uncertainty instigates a process of "requalification" through "groping," as
Dewey (1986:191–192) called it, involving the identification of the limits or
boundaries of an indeterminate situation . This is comparable to the diffuse
writhing and wiggling movements that McGraw found infants making in
their initial attempt to explore the limits of their immediate environment.

The next stage of judgment involves the effort to distinguish specific characteristics and kinds against the background of more general phenomena.
Generic propositions are invoked at this point, according to Dewey, to distinguish traits according to whether they belong to one kind or another
(Dewey 1986:193–194). McGraw's infants illustrated this stage in their deliberate efforts to attain balance by controlling the frequency and direction
of their movements, contributing to one form of locomotion rather than another. Finally, a point is reached after experimental operations have been
performed as specified by "universal propositions" when extraneous elements (i.e., excessive or unnecessary steps or hypotheses proven incorrect)
are eliminated and those remaining, sufficiently integrated (as exemplified
by a stable pattern of integrated behavior) so that final warranted judgments
can be embodied in propositional form. Dewey (1986:194–195) characterizes this stage of inquiry as a "reaction from some into all," when phenomena sharing the same qualitative attributes as one kind are integrated to form
generalizations applying to all such kinds.

McGraw's studies of infant locomotion furnished the evidence Dewey sought to argue that the focus and integrity of inquiry, like the equilibrium of the organism, is maintained despite the ever changing dimensions of the problem under study. Consequently, the indeterminacy of a series of seemingly separate and distinct events or occurrences can be overcome by showing that they are connected through some more basic underlying rhythm or integral pattern. By assigning interaction this functional role in inquiry, Dewey (1986:452–453) demonstrated how facts become interrelated or correlated through the positions they occupy within a whole series or pattern of propositions, giving inquiry a "cumulative force" towards a unified conclusion. Thus successful inquiry, like erect locomotion, according to Dewey, is demonstrated by whether the momentum from the measured steps taken to understand and overcome a problem carries us forward to face new uncertainties without loosing our footing or breaking stride.

McGraw (1935:5–8) advanced a philosophy of science in her book, *Growth* (see Chapter 9), based on the analogy between the stages of theory building in psychology and erect locomotion that anticipated Dewey's arguments in *Logic*. McGraw describes how infants engaged in reflex stepping show little awareness until they are involved in the preliminary stages of walking. At that time, their movements seem excessive and uncoordinated, swinging from one extreme to another as typified in the adoption of a wide stance followed by a narrow one. Eventually however, excess activity is reduced and a well-coordinated and stable, integrated gait is attained.

By analogy, McGraw (1935:5) observes that, in centuries past, scientists made little conscious effort to obtain information about child development. Once systematic studies were launched at the turn of the century, according to McGraw, a lot of energy was wasted without much forward advance. Then a dominant idea or theory emerged such as Behaviorism that McGraw says exhibited a characteristic "tendency to exaggerations in its claims" until checked by a competing theory such as Gesell's maturationism. This process of alternation between competing hypotheses, according to McGraw (1935:5), reaches a new plateau when "the essence of the two may become crystallized into an unchallengeable truth, a specific, well-determined principle which can stand alone and walk through the universe of ideas fullgrown."

Similarly, Dewey (1986:197-199) argued in *Logic*, contrary to Thomas Kuhn (1972), that science advances not simply by revolutionary bursts of insight, but through a gradual process of expansion and contraction of generalizations to attain a better fit between hypotheses and evidence. New theories often contradict but rarely completely overthrow older theories. While Dewey believed that science advances by overcoming anomalies by anticipating the unexpected, he also contended that new insights are yielded from older generalizations when they are used as analogies to understand a problem from a new perspective. Consequently, scientific consensus occurs when competing lines of inquiry identify *mutual* problems that make possible unified solutions. As Leahey (1992) argues, scientific progress in psychology may not have occurred through paradigm shifts, as some argue, but through a continuous remodeling of assumptions and methods, enabling dominant perspectives such as behaviorism to assume new forms.

The Ontogeny of Growth
and Consciousness

What set McGraw's research apart most from that of her contemporaries was her attempt to discover the mechanisms of growth governing development (see Introduction to Part 3) and determine how consciousness (i.e., awareness) contributes to the continuous expansion and integration of behavior. These were ambitious goals during an era in which developmental biology was in its infancy (see Oppenheimer 1966) and the issue of consciousness was on the verge of extinction in a research field dominated by behaviorism. McGraw's essays in Part 4, particularly Chapter 14, "Maturation of Behavior," presents a comprehensive review of evolutionary theories and techniques, including embryological methods available for the study of neurobehavioral development when she undertook her research in the 1930s. As Oppenheim (1982; 1992) points out, contemporary researchers have not fully appreciated the contributions of pioneering neuroembryologists such as Hans Spemann, Ross Harrison, Victor Hamburger, and George Coghill, who strongly influenced studies of behavioral ontogeny undertaken by McGraw, Gesell, and other psychologists.

Stephen J. Gould, a paleobiologist, has rekindled an interest in human evolution with numerous popular essays reconstructing and updating controversial issues about ontogeny debated by pioneering nineteenth century scientists such as Charles Darwin, Thomas Huxley, and Samuel Butler among others. Gould (1987) argues that flexibility constitutes our most important human attribute because it has enabled us to evolve behaviors for successful adaptation. Gould (1987:63) believes that we owe this flexibility to our large brains which have been acquired through an evolutionary process of *neoteny* or the slowing down of development, so that we retain in adulthood the juvenile traits of chimpanzees, our closest ancestors. Neoteny is best illustrated, according to Gould, by the retention of the capacity for rapid neural growth characteristic of fetal stages and by the relatively longer duration of maturation and learning processes. Gould (1977) asserts on the basis of the paleontological record that humans are "neotenic" apes.

In their pathbreaking work on the evolution of ontogeny, McKinney and McNamara (1991:292–295; 314) dispute Gould's hypothesis. They persuasively demonstrate that the human acquisition of new functional capacities could be explained by principles of discontinuous growth (i.e., changes in growth rate affect differently the form, sequence, and function of traits) illustrated in McGraw's pioneering research. They contend that human evolution may not be best explained by neoteny (i.e., delay of *onset* due to slower growth) or prolonging into adulthood the youthful attributes of ancestors, as Steven Jay Gould (1977) supposes. They argue instead that development occurs through a process of hypermorphosis or the delayed *offset* of growth due to selective acceleration and deceleration of separate growth processes. This multivalent process increases the chance for variation by enabling the brain to grow larger and the body to acquire capabilities beyond those possessed by ancestors.

In this regard, McGraw showed that the integration of motor skills is due not only to the delayed onset of cortical control but to the delayed offset of phyletic (older) traits, prolonging their influence in behavioral development. For example, McGraw and Breeze (1941:298–299) found that an infant does not attain a stable gait immediately, but must gradually master several interrelated components of bilateral walking, foreshadowed in the supported exercise of stepping reflexes. She also discovered that these so-called stereotypical reflexive reactions could be stimulated to increase learning through accelerated motor development, as demonstrated in Johnny's numerous physical feats considered impossible at his age. Nor did McGraw's research lend any credence to the fear that training missed is learning foregone forever. Critical periods of growth in early childhood provide opportune times for acquiring a skill easily and smoothly but do not make acquisition impossible at a later time. New skills can be added just as old habits can be changed when a new posture and attitude is adopted, bringing about a new awareness of the contingency of sequence and possibility of change.

McKinney and McNamara contend that superficial anatomical and developmental resemblances between chimpanzees and humans are misleading and argue that humans cannot be distinguished from apes simply by their larger brains and relatively slower rates of maturation. Instead, they believe that the essential difference between species is that humans develop for a longer time in each growth phase, with each phase progressing at different rates. The cumulative effect of delayed offset and variable phase rate, according to them, "stretches out" development, producing significant differences in species with similar developmental sequences (McKinney and McNamara 1991:293). For example, distinctive human traits associated with intelligent behavior emerge during late ontogeny when chimpanzee development has already been completed (McKinney and McNamara 1991: 314–315). Some scientists speculate that changes in the pattern of human development originated in structural changes in the brain that accelerated the rate and extended the phases of neural growth, vastly increasing the interneuronal connections available to support complex behavior (Purves 1988). Thus the kinetic energy diverted from direct expenditure during prolonged periods of growth provides an enormous reservoir of potential energy available for the redevelopment and expansion of human capacities.

McGraw (1934) had the occasion to personally observe chimpanzee behavior when she and Dewey visited the Yerkes Anthropoid Station in Orange Park Florida in early 1935 (see Yerkes 1935). McGraw (1941b) formulated her impressions from her two week session in a remarkable unpublished paper that speculates on a pivotal event leading chimpanzees and humans down separate evolutionary paths. McGraw was struck by the appearance in infant chimpanzees of what seemed to be vestiges of erect locomotion. For example, baby chimps assumed sitting postures, with legs extended outward, and with palms flat (rather than resting on knuckles) preparatory to rising to a standing position. They also simulated the beginning phase of erect locomotion with arms in extension and abduction and engaged in bipedal walking much more frequently than adults. This suggested an astonishing possibility, according to McGraw (1941b:2), "that those forces which led to the specialization of the Chimpanzee in the phylogenetic sequence are actually younger than the trend toward bipedal progression and when they emerged they were so forceful that they inhibited the complete evolution of bipedal progression in that species," adding that, "the anlage (i.e., functional precursor) of bipedal progression, it would seem, actually emerged before the specialization by the chimpanzee."

While tantalizing, the possibility that chimpanzees may represent the parallel or even subsequent but unsuccessful attempt by a species closely related to human ancestors to master erect locomotion must await further paleontological evidence. However, it seems likely that the forces favoring bipedalism were more complex than conventional environmental hypotheses which suggest that erect locomotion was an adaptive response to life on the savannah plains. For example, the chimpanzee's rudimentary ability to find and use tools may have enhanced their foraging skills in forests, thus reducing pressure for new food sources in grassland environments. Plooij (1984) has demonstrated that chimpanzee babies exhibit a remarkable similarity to human infants in sequential patterns of development. Nevertheless, chimpanzees fail to attain the level of sensori-motor integration near the end of their second year when most infants can manipulate objects and walk. This suggests the possibility that at some crucial point in their ontogeny, chimpanzees failed to develop neuroanatomical structures and interneuronal connections (i.e., between cerebellum and cortex) to sustain the balance necessary to free their hands completely for inquiry and learning.

The pace of technological change and disciplinary integration in developmental biology has made possible innovative research on growth and developmental processes that McGraw only imagined. Neurobiologists increasingly collaborate with immunologists, biochemists, and geneticists using cytometric, recombinant, and transcriptional methods to understand the lineage, sequence, and timing of molecular and cellular events governing the origin, structures, and functions of living organisms (Brown 1993:4–5). For example, Ronald Oppenheim, a neurobiologist noted for his research on nerve growth factors and Gerald Edelman, a Nobel laureate recognized for important discoveries in immunological processes, have collaborated with other colleagues in research to identify molecular agents responsible for the adhesion of neural cells necessary to attain functional specificity (see Shiga, et al.1993).

Advances in electroencephalography (i.e., study of brain waves) and use of non-invasive ultrasound to study fetal growth also promise to revolutionize our understanding of the processes and patterns of early growth and development. Neurobehavioral research will also benefit from applying computer assisted morphometric techniques, pioneered by Bookstein (1991), to determine how transformations in neuroanatomical structures occur in ontogeny and identify their effect on functional development. By employing these and other techniques, scientists of developmental phenomena will reveal whether growth processes possess common attributes at the molecular, cellular, neural, and behavioral level, and explain how they interact during development. Research of this kind will infuse on-going theoretical debates about ontogeny with a concreteness now missing in these largely speculative, academic disputes.

McGraw's research also rubbed against the grain of behaviorist orthodoxy which reduced consciousness to a minor epiphenomenon of behavior. Behaviorists ruled out attitudes as reliable indicators of consciousness since they were thought to be either hereditary expressions of temperament or instinctual (or defensive) reactions triggered by fear or uncertainty. However, McGraw found evidence that infant's express their attitudes through gestures and believed that this provided objective evidence of an infant's awareness of his or her own feelings in relation to external conditions impinging on its welfare. She also discovered that memories of early motor experiences could be subsequently recalled when children found appropriate terms to express them linguistically (McGraw 1939; 1942).

Gerald Edelman's (1989) seminal research on the evolution and development of the human brain underscores the importance to human intelligence of a prelinguistic memory, and the capacity to retrieve and re-process fragmented information into new integrated forms. Edelman (1989:126;132) contends that the human brain is preeminently an organ of succession in the sense that higher order consciousness and learning depend on the brain's capacity to track successive sensori-motor states. He hypothesizes that sequential motor behavior makes possible recategorization through memory and the recombination of ideas or movements essential to adaptation and individual transformation. Edelman suggests how the cerebellum and cortex work in conjunction to "categorize. ... a smooth succession of motions in gestures and the succession of gestures in synergies" to make learning possible. Significantly, according to Edelman, memory serves as the primary mechanism for "re-entrant signaling," or the re-processing of information into usable forms that allows us to continuously alter our actions by reordering the sequence of previous activities. This mnemonic basis of consciousness is aptly captured by the title of Edelman's book, *The Remembered Present*.

Edelman (1989:142) hypothesizes that acquisition of language did not require major changes in the brain but simply involved the transfer of functions from cerebellum to cortex. Accordingly, concepts embody prelinguistic "judgments made about classification of stimuli long past." Edelman clarifies the evolutionary implications of this assertion by saying that:

> the ability to have concepts must have required the evolution of brain regions that are capable of these more recently evolved functions but are nonetheless structurally based on the earlier functions of perceptual categorization, memory and learning. However, it is not necessarily true that any basically new *microstructure* of the brain had to be developed in order for such a capacity to emerge. One can reasonably suppose that the frontal, parietal, and temporal cortex and basil ganglia are good candidates to serve as bases for temporally delayed judgments and actions entailed by the having of concepts and necessary for their generation. If this is true, lobsters and perhaps even birds do not have concepts, but dogs might (Edelman 1989:143).

McGraw's follow-up study of Johnny and Jimmy (see Part 2, Chapter 3) supports Edelman's contention about the role of succession in memory and the importance of pre-linguistic judgment in learning. Edelman's theory is well illustrated in an amusing anecdote in McGraw's follow-up study, involving Johnny's reactions when he returned to Babies Hospital and Neurological Institute over two years after the initial experiments were completed. McGraw recalled that Johnny would normally skate through the tunnel from Babies Hospital to the Neurological Institute to get to the swimming pool. When he reentered the tunnel again at 43 months old, his "eyes widened and he suddenly remarked with a sweeping gesture," according to McGraw (1939:18), 'This is where we go skating!' On another visit to the tunnel a year later, Johnny asked, 'Where's the bathroom?' When told that he would have to wait until he returned to the laboratory, he replied, 'Oh there's one way, way down there,' pointing in the direction of Neurological Institute. It was then noted that when Johnny was seventeen months old, he was being trained in toilet habits, and as soon as he arrived at the Neurological Institute he was rushed to the toilet."

McGraw (1939:18) commented on these two remarkable instances of Johnny's "recall through remote memory" by saying that:

This recall is interesting beyond the mere indication of memory for early experiences. At the time the tunnel journeys were ended Johnny had no verbal means of expressing the act of skating. He recognized skates by name but he had no word in his own vocabulary to indicate them. Here his remote memory is entirely integrated with verbal expression which was acquired at a later date. ... If the direct memory of experiences during the first two years of life are as indelible as these incidents indicate, then the weight attributed to influences imposed upon the infant and young child gains significance.

These episodes involving childhood memories are significant because Johnny's recall of his experience was triggered in part by remembering a particular event within the sequence of activities in which it occurred. Johnny employed sequential memory in his subsequent visit to find a shorter route through the tunnel to get to the bathroom at the Neurological Institute. Moreover, the acquisition of new concepts between visits enabled Johnny to find new meaning (or connections in the sequel) associated with skating in the tunnel through terms unavailable to him before. Therefore, language provided a powerful tool for retrieving a previous experience, as Edelman argues (i.e., through recategorization and resequencing) that had been retained on the basis of an overall impression, an awareness, or neuromuscular feeling associated with it.

CONTINUITY AND VARIABILITY
IN DEVELOPMENT

McGraw established an innovative program of laboratory instruction designed to analyze pre-verbal signals of infant development at Briarcliff (a women's college that closed in 1974) where she taught for almost two decades from 1953 through early 1972. In his introduction to Part 4, Donald Dewsbury thoughtfully examines the premises and central themes underpinning McGraw's psychobiological conception of human development, including an essay dating from her Briarcliff experience. McGraw (1969; 1971) recounted setting up the infant laboratory as a pedagogical device (see Chapter 15) so that students could cultivate an "intuitive sensitivity" to infant development by correctly interpreting their pre-linguistic "signals" of growth. Accordingly, each student was assigned to observe for 2 hours a week a newborn baby or one considerably under 2 years of age. Students were responsible for submitting video tapes and detailed reports that were subsequently discussed by the whole class. Not only were infant behaviors analyzed but future behaviors were predicted, based on existing trends and emerging capabilities. Consequently, students could better evaluate whether they had properly interpreted growth signals and accurately gauged infant progress. One former student remarked that:

What she [McGraw] prepared me for in teaching was to be able to take any parent and stand up and say this is why I did it, and I did it because I can prove this or that aspect of growth and so forth, and be able to take the criticism, think about it and deal with it, and be constructive in feeding back what it was that they were seeing or to learn from it so we can improve ourselves. ... What she taught me through experiential learning was a very able way of dealing with conflict, to think ahead when I did something to make sure that I knew why I was doing it. ... McGraw had developed the curriculum very thoughtfully to deal, I think, with many things we needed to learn at that maturation level where we were and the skills we needed to develop to go on to do the things we needed to do as individuals, as women, and also as teachers. ...I have been a teacher first and foremost. But it was the children stimulating me which caused me to get ideas and then feed back into them and into my own children (Harter 1992).

Perhaps it comes as no surprise that McGraw had drafted a "Mothers Manual" nearly two decades before she got her laboratory underway at Briarcliff. Her manual included a protocol for observations and tests of motor development conducted by her students. McGraw attempted unsuccessfully to publish the manual in the late 1930s as a popular guide for parents and thus, would have predated Dr. Benjamin Spock's (1946) enormously successful book on baby care by several years (See Chapter 4, "Signals of Growth," for the nearest approximation of format and style of the manual). McGraw's manual was motivated by her desire to restore self-confidence in child rearing that she believed Freudian psychology had taken from parents, by encumbering them with deep-seated anxieties and feelings of inadequacy about the love and care for their children (see Chapters 1 and 16).

McGraw did not share her colleagues' awe and curiosity for psychoanalysis and her negative attitudes probably were responsible for her stormy internship at the Columbia Institute for Child Guidance headed by Freudians David Levy and Lawson Lowrey (McGraw 1972a:4–5). Years later, McGraw explained her reasons in an unpublished paper for rejecting Freud's theories about childhood by saying:

There were several reasons for my reluctance. Freud had not experimented with infants and I could not identify his assertions about infancy with the babies I had under observation. He promulgated his ideas about infancy from his analysis of patients. I considered his evocations in that area as armchair philosophy. Funny, I never once thought of 'JD' [John Dewey] as an armchair philosopher. He wrote many erudite, difficult books, without building up a nomenclature of his own. Because of his ability to see the internal ramifications of concepts, he saw depth in the meaning which the ordinary reader does not quickly understand nor grasp (McGraw 1980:63).

McGraw was not opposed to exploring the clinical implications of growth processes, despite her misgivings about Freudian psychology (including

An infant tests the controls on a "Child Technology" machine built by McGraw's husband, Rudy Mallina, for use in her laboratory at Briarcliff College, 1969. (Photo by Robert T. Propper, courtesy of Mitzi Wertheim.)

Gesell's age-based norms). She believed that clinical judgments should be drawn cautiously and be based on a complete understanding of the complex interdependence of neural, behavioral, and environmental factors in development.

For example, McGraw had the opportunity to compare notes with Lewis P. Lipsitt (1979) in the late 1970s regarding the timing and potential factors contributing to sudden infant death syndrome. Lipsitt (1979) contends that the period in early infancy (between 2 and 4 months) when transient reflex functions are waning with the onset of cortical control, as McGraw demonstrated, poses for some infants a special period of jeopardy and risk. Sudden infant death syndrome (and perhaps, the failure to thrive), whose symptoms include inadequate postural control and respiratory occlusion, may be explained, according to Burns and Lipsitt (1991) by possible deficits in neural functioning combined with deficient environmental supports (i.e., motor stimulation) compromising the ability of these infants to fully develop adaptive behaviors. Although withholding judgment as to specific causes of sudden infant death, McGraw concurred with Lipsitt that the period of special risk coincided with a critical period during infancy when the attainment of balance was instrumental to the mastery of motor activities such as sitting up and crawling.

Touwen (1976; and see his afterword) took seriously McGraw's reservations that our knowledge about abnormal functions depends crucially on how well we understand *normal* developmental processes. Touwen's (1971; 1976) research corroborates McGraw's discovery that normal development is marked by extensive variability in the timing and attainment of motor functions. Touwen (1978; 1994) contends that it is the *capacity to vary* rather than the ability to approximate a norm, as Gesell proposed, that has primary diagnostic significance for development. Touwen indicates in his afterword "that the lack of ability to vary will inevitably lead to a lack of adaptive (both active and passive) capacities," singling out cerebral palsy which, Touwen observes, involves "the lack of sufficient and proper strategies" to leading "non-adaptive and stereotyped motor behavior." Using variability as the point of departure is likely to refocus research on factors involved in growth and developmental processes at every level (i.e., molecular, cellular, physiological, and so forth) that allow for the enormous range of human thought and behavior despite the relative outward similarity in human form and functions.

LETTING BABIES BE OUR TEACHERS

A final observation about McGraw's rhetorical style is appropriate because of its singular contribution to understanding why babies may be our best teachers about growth and learning. McGraw grew up in an era governed by the general rule that children should be seen but not heard. Development by disciplined silence, however, proved to be an unrealistic and ineffective method of child-rearing as witnessed by the alarming rise in juvenile delinquency during the 1920s. A huge industry grew up around the use of standardized tests of I.Q. and other psychological batteries to differentiate children according to aptitude and social adjustment. Criminologists and psychiatrists turned to Freud and Watson to find ways to hear, interrogate, and modify the voices muted in early childhood so that a turbulent adolescence might be prevented (Snodgrass 1981). Dr. Spock's (1968) popular pediatric guide to child-rearing enhanced the legitimacy and prestige of physicians, psychologists, and other professionals upon whom parents grew ever more dependent for advice.

McGraw sensed that what was missing in these well intentioned forms of professional assistance was a willingness to watch and listen to infants and children without preconceptions about the meaning of their behavior.

McGraw's ability to communicate effectively non-verbally with infants explains perhaps why there is such a palpable feeling of mutual understanding in the photographs of her with infants appearing in this book. McGraw believed that psychoanalytic and behaviorist theories advanced over-schemetized conceptions of child development that offered conflicting prescriptions for child-rearing. She also argued that standardized tests of intelligence put undue emphasis on achievement that obscured the inventive processes involved. McGraw's critical attitudes were reflected in an impertinent, mocking, and deprecating style adopted in essays written for a popular audience. Her frequent belittling of dominant psychological theories in these essays (e. g., "Signals of Growth") was done to call attention to the errors of dichotomous reasoning from either nature or nurture. Although appealing, this tactical style accentuated the shortcomings of and differences between opposing theories. Nevertheless, McGraw made a concerted effort in professional journals (see Part 4) to outline collaborative research strategies and to address theoretical controversies.

McGraw's statement that best exemplifies her commitment to interdisciplinary collaboration was one she shared with Philip Zelazo when she said:

> ... one of the most urgent areas for systematic research involves the study of the psychoneural communications between the cortical and subcortical areas of the brain—the communications and 'feed–back systems between them—in all activities common to the growing infant. Maybe some of the younger experimentalists can be inspired to undertake it in collaboration with neurologists. The electroencephalogram can yield more reliable data than were available to us in the thirties (McGraw 1972b).

McGraw believed that the heavily prescriptive popular literature on child-rearing missed two of the most salient features of developmental processes: that learning and growth occur through a reciprocal process of communication, and that uncertainty enlarges the realm of judgment. The inability to talk does not mean that an infant or young child cannot communicate or reason with an adult. McGraw thought parents fully capable of interpreting infant behavior and pre-verbal gestures, or "signals of growth," and providing stimulation during "critical periods" for children to attain their maximum potential during development. Growth is also open-ended and unpredictable. It is subject to sudden fluctuations in tempo, focus, and direction. The very energy and dynamism intrinsic to growth that McGraw believed stimulates awareness also provokes doubt and uncertainty. But the need for judgment arises precisely in situations of ambiguity, giving vent to exploration to define the limits and direction of potential action. McGraw believed that through communication parents and children could get on the same wave length or frequency to make learning a shared experience.

McGraw's life took a fateful turn when she struck up a correspondence with John Dewey after first reading about him in an article in the *Independent* titled: "John Dewey: Teacher of Teachers." Dewey's influence on American educators remains unrivaled, but we have yet to fully fathom the lessons that Myrtle McGraw's babies can teach us if we only let *them* be our teachers!

References

Antler, J. 1982. "Progressive Education and the Scientific Study of the Child: An Analysis of the Bureau of Educational Experiments." *Teachers College Record* 83:559–591.

Bergenn, V.W., T.C. Dalton, and L.P. Lipsitt. 1992. "Myrtle B. McGraw: A Growth Scientist." *Developmental Psychology* 28:381–389.

Bookstein, F.L. 1991. *Morphometric Tools for Landmark Data: Geometry and Biology.* New York: Cambridge University Press.

Brown, D. 1993. "Developmental Biology has Come of Age." In *Evolutionary Conservation of Developmental Mechanisms.* ed. A.C. Spradling. New York: Wiley-Liss.

Burns, B. and L.P. Lipsitt. 1991. "Behavioral Factors in Crib Death: Toward an Understanding of the Sudden Infant Death Syndrome." *Journal of Applied Developmental Psychology* 12:159–184.

Burks, A.W., ed. 1958. *Collected Papers of C. S. Peirce.* Cambridge, Mass.: Harvard University Press.

Child, C.M. 1921. *The Origin and Development of the Nervous System.* Chicago: University of Chicago Press.

Coghill, G.E. 1933a. "The Neuro-Embryologic Study of Behavior: Principles, Perspectives, and Aims." *Science* 78:131–136.

———. 1933b. Letter to C.J. Herrick. (October 27). Neurology Collection, C. Judson Herrick Papers. Spencer Research Library, University of Kansas, Lawrence, Kansas.

———. 1934. Letter to A. Gesell. (April 23). Arnold Gesell Papers, Rare Books and Manuscripts, Library of Congress, Washington, D.C.

Cravens, H. 1993. *Before Head Start: The Iowa Station and America's Children.* Chapel Hill, NC: University of North Carolina Press.

Crowell, D.H. 1967. "Infant Motor Development." In *Infancy and Early Childhood,* ed. Y. Brackbill. New York: Free Press.

Dalton, T.C. 1994. "Challenging the Group Bias of American Culture." Review of *Before Head Start: The Iowa Station and America's Children* by Hamilton Cravens. *Contemporary Psychology* forthcoming.

Dalton, T.C. and V.W. Bergenn. 1994. "John Dewey, Myrtle McGraw, and the *Logic*: An Unusual Collaboration in the 1930s." forthcoming.

Dennis, P. 1989. "'Johnny's a Gentleman but Jimmie's a Mug': Press Coverage During the 1930s of Myrtle McGraw's Study of Johnny and Jimmy Woods." *Journal of the History of the Behavioral Sciences* 25:656–370.

Dewey, E. 1935. *Behavior Development in Infants: A Survey of the Literature on Prenatal and Postnatal Activity, 1920-1924.* New York: Columbia University Press. (reprint, New York: Arno, 1972).

Dewey, J. 1930. Letter to C. Chisholm. (February 28). Rare Books and Manuscripts, Butler Library, Columbia University, New York.

_____. 1934a. Letter to M. McGraw. (May 31). Center for Dewey Studies, Southern Illinois University, Carbondale, Illinois.

_____. 1935a. "Introduction," In *Growth: A Study of Johnny and Jimmy.* M. McGraw. New York: Appleton Century. (reprint, Arno Press, 1972)

_____. 1935b. Letter to A. Bentley. (July 26). Arthur Bentley Papers. Manuscripts Department. Lilly Library, Indiana University, Bloomington, Indiana.

_____. 1976a. "Principles of Mental Development as Illustrated in Early Infancy." In *John Dewey: The Early Works, Vol.1: 1899-1901,* ed. J.A. Boydston. Carbondale, Il: Southern Illinois University Press.

_____. 1976b. "The Evolutionary Method as Applied to Morality: I. Its Scientific Necessity." In *John Dewey: The Middle Works, Vol 2: 1902-1903.* ed. J.A. Boydston. Carbondale, Illinois, Southern Illinois University Press.

_____. 1981. "Experience and Nature." In *John Dewey: The Later Works, Vol. 1: 1925-1953,* ed. J.A. Boydston. Carbondale, Illinois: Southern Illinois University Press.

_____. 1986. "Logic: The Theory of Inquiry." In *John Dewey: The Later Works, 1925-1953, Vol. 12,* ed. J.A. Boydston. Carbondale, Illinois: Southern Illinois University Press.

_____. 1988a. "The Unity of the Human Being." In *John Dewey: The Later Works, 1938-1939, Vol. 13,* ed. J.A. Boydston. Carbondale, Illinois: Southern Illinois University Press. 1986.

_____. 1988b. "Human Nature and Conduct." In *John Dewey: The Middle Works, Vol 2: 1902-1903,* ed. J.A. Boydston. Carbondale, Illinois, Southern Illinois University Press.

Edelman, G.R. 1989. *The Remembered Present: A Biological Theory of Consciousness.* New York: Basic Books.

Elsberg, C. 1944. *The Story of a Hospital: The Neurological Institute of New York, 1909-1938.* New York: Paul F. Hoeber.

Fowler, W. 1983. *Potentials of Childhood: Vol. 1, A Historical View of Early Experience.* Lexington MA: Lexington Books.

Frank, L.K. 1933. Memorandum of an Interview with Myrtle McGraw. (March 13). General Education Board, Record Group I., Series 1.3, Box 370, Folder 3858. Rockefeller Archive Center, Tarrytown, New York.

_____. 1935a. "Structure, Function, and Growth," *Philosophy of Science* 2:210–235.

_____. 1935b. "The Problem of Child Development." *Child Development* 1:7–18

_____. 1949. *Society as the Patient.* New Brunswick New Jersey: Rutgers University Press.

_____. 1962. "The Beginnings of Child Development and Family Life Education in the Twentieth Century." *Merrill-Palmer Quarterly of Behavior and Development* 8(4):207–228.

Gesell, A. 1934. *Infant Behavior: Its Genesis and Growth.* New York: McGraw-Hill.

———. 1939. "Reciprocal Interweaving in Neuromotor Development," *The Journal of Comparative Neurology* 10:161–180.

Gottlieb, G. 1991. "Experiential Canalization of Behavioral Development: Theory." *Developmental Psychology* 27:4–13.

Gould, S.J. 1977. *The Mismeasure of Man.* New York: W.W. Norton.

———. 1987. *An Urchin in the Storm.* New York: W.W. Norton.

Harter, N. 1992. Interview by T.C. Dalton. (October 14). Bethesda, Maryland.

Hetherington, E.M. and Parke, R.D. 1993. *Child Psychology* 4th ed.,New York: McGraw Hill.

Herrick, C.J. 1910a. "The Relations of the Central and Peripheral Nervous Systems in Phylogeny." *Anatomical Record* 36:645–652.

———. 1910b. "The Evolution of Intelligence and its Origins." *Science* 31: 7–18.

———. 1913. "Some Reflections on the Origin and Significance of the Cerebral Cortex." *The Journal of Animal Behavior* 3:222–236.

———. 1949. *The Evolution of Human Nature.* Chicago: University of Chicago Press.

———. 1962. *Neurological Foundations of Animal Behavior.* New York: Holt (reprint, New York: Hafner, 1924).

Heyl, K.A. 1989. Telephone Interview by T.C. Dalton. (March 13). Norwich, Vermont.

Johnson, B. 1925. *Mental Growth of Children in the Relation to the Growth in Bodily Development.* New York: Dutton.

Kast, L. 1936. *A Review by the President of Activities for the Six Years Ending December 31, 1936.* New York: Josiah Macy Jr. Foundation.

Kingsland, S. 1990. "Toward a Natural History of the Human Psyche: Charles Manning Child, Charles Judson Herrick, and the Dynamic View of the Individual at the University of Chicago." In *The Expansion of American Biology,* eds. K.R.Benson, J. Manschein, and R. Rainger. New Brunswick: Rutgers University Press.

———. 1993. "A Humanistic Science: Charles Judson Herrick and the Struggle for Psychobiology at the University of Chicago." *Perspectives on Science* 1:1–33.

Kuhn, T. 1972. *The Structure of Scientific Revolutions.* Chicago: University of Chicago.

Leahey, T.H. 1992. "The Mythical Revolutions of American Psychology." *American Psychologist* 47:308–318.

Lipsitt, L.P. 1979. "Critical Conditions in Infancy: A Psychological Perspective." *American Psychologist* 14:973–980.

———. 1990. "Myrtle B. McGraw (1899–1988)" *American Psychologist* 45:977.

McGraw, M.B. 1926. Letter to A. Gesell. (March 19). Arnold Gesell Papers, Manuscript Division, Library of Congress, Washington, D.C.

———. 1927. Letter to A.Gesell. (March 2). Arnold Gesell Papers, Manuscript Division, Library of Congress, Washington, D.C.

———. 1932. Letter to C.J. Herrick. (December 9). Neurology Collection, C. Judson Herrick Papers. Spencer Research Library, University of Kansas, Lawrence Kansas.

_____.1934. Letter to L.K. Frank. (December 21). General Education Board, Record Group 1.3, Box 370, Folder 3858, Rockefeller Archive Center, Tarrytown, New York.

_____. 1935. *Growth: A Study of Johnny and Jimmy.* New York: Appleton-Century, 1935 (reprint, New York, Arno Press, 1975).

_____. 1939. "Later Development of Children Specially Trained During Infancy: Johnny and Jimmy at School Age." *Child Development* 10:1–19.

_____. 1941a. "Development of Neuromuscular Mechanisms as Reflected in the Crawling and Creeping Behavior of the Human Infant." *Journal of Genetic Psychology* 58:83–111.

_____. 1941b. "A Thought on the Process of Evolution." Unpublished manuscript. Myrtle B. McGraw Papers, Special Collections, Millbank Memorial Library, Teachers College, Columbia University, New York.

_____. 1942. "Appraising Test Responses of Infants and Young Children." *Journal of Psychology* 14:89–100.

_____. 1943. *The Neuromuscular Maturation of the Human Infant.* New York: Columbia University Press.

_____. 1967. Interview with K. Duckett. (February 9). Hastings-on Hudson, New York. Special Collections, Morris Library, Southern Illinois University, Carbondale, Illinois.

_____. 1969. "An Open Letter to Parents of Young Infants." Unpublished manuscript, Myrtle B. McGraw Papers, Special Collections, Millbank Memorial Library, Teachers College, Columbia University, New York.

_____. 1971. "A Laboratory Approach to Training Undergraduates in Child Development." Unpublished manuscript, Myrtle B. McGraw Papers, Special Collections, Millbank Memorial Library, Teachers College, Columbia University, New York.

_____. 1972a. Interview with M. Senn. (May 9). Milton Senn Collection. History of Medicine Division, National Library of Medicine, Bethesda, Maryland.

_____. 1972b. Letter to P. Zelazo. (June 3). Myrtle B. McGraw Papers, Special Collections, Millbank Memorial Library, Teachers College, Columbia University, New York.

_____. 1979. Letter to R.W. Oppenheim. (December 31). Myrtle B. McGraw Papers, Special Collections, Millbank Memorial Library, Teachers College, Columbia University, New York.

_____. 1980. "Growing Up With and Without Psychology." Unpublished manuscript, Myrtle B. McGraw Papers, Special Collections, Millbank Memorial Library, Teachers College, Columbia University, New York.

_____. 1985a. "Professional and Personal Blunders in Child Development Research." *Psychological Record* 35:165–170.

_____. 1985b. "Thoughtways" Lecture at *Conference on Motor Development* Teachers College. (November 1). New York City.

_____. 1990. "Memories, Deliberate Recall, and Speculation." *American Psychologist* 45:934–937.

McGraw, M.B. and K.W. Breeze. 1941. "Quantitative Studies in the Development of Erect Locomotion." *Child Development* 12(3):267–235.

McKinney, M.J. and K.J. McNamara. 1991. *Heterochrony: The Evolution of Ontogeny.* New York: Plenum.

Merriam, J. 1933. Letter to F. Keppel. (March 15). The Normal Child Development Study, Carnegie Corporation Papers, Rare Books and Manuscripts, Butler Library, Columbia University, New York.

Mitchell, L.S. 1925. Letter to Trustees. (May 6). Laura Spelman Rockefeller Memorial Record Group I, Series 3.5, Box 27, Folder 279, Rockefeller Archive Center, Tarrytown, New York.

Nagel, E. 1977. "Can Logic be Divorced from Ontology?" In *Dewey and His Critics.* ed. S. Morgenbesser. New York: Journal of Philosophy.

_____. 1986. "Introduction." In *John Dewey: The Later Works, 1925-1953, Vol. 12,* ed. J.A. Boydston. Carbondale, Illinois: Southern Illinois University Press.

New York Times. 1935 (January 16):19

Oppenheim, R.W. 1982. "The Neuroembryological Study of Behavior: Progress, Problems, Perspectives." In *Current Topics in Developmental Biology, Vol. 17. Part 3. Neural Development,* ed. R.K. Hunt. New York: Academic Press.

_____. 1992. "Pathways in the Emergence of Developmental Neuroethology: Antecedents to Current Views of Neurobehavioral Ontogeny." *Journal of Neurobiology* 23:1370–1403.

Oppenheimer, J. 1966. "The Growth and Development of Developmental Biology." In *Major Problems in Developmental Biology,* ed. J. Locke. New York: Academic Press.

Plooij, F.X. 1984. *The Behavioral Development of Free Living Chipanzees.* Norwood, NJ.: Ablex

Pool, L.J. 1975. *The Neurological Institute of New York.* Lakeville, CT: Pocket Knife Press.

Prechtl, H F.R. 1989. "Fetal Behavior." In *Fetal Neurology,* ed. A. Hill and J.J. Volpe. New York: Raven Press.

Purves, D. 1988. *Body and Brain: A Trophic Theory of Neural Connections.* Cambridge: Harvard.

Rappleye, W. 1935. Letter to L.K. Frank. (December 2). General Education Board, Record Group I., Series 1.3, Box 370, Folder 3858. Rockefeller Archive Center, Tarrytown, New York.

Rappleye, W. 1955. *The Josiah Macy Jr. Foundation: Twentieth Anniversary Review, 1930-1955.* New York: Josiah Macy Jr. Foundation.

Russell, B. 1977. "A Critique of Professor Dewey's *Essays in Experimental Logic*" In *Dewey and His Critics,* ed. S. Morgenbesser. New York: Journal of Philosophy.

Shiga, T., T. Shirai, M. Grumet, G. Edelman, and R. Oppenheim. 1993. "Differential Expression of Neuron-Glia Adhesion Molecule (Ng-Cam) on Developing Axons and Growth Cones of Interneurons in the Chick Embryo Spinal Cord: An Immunoelectron Microscopic Study." *Journal of Comparative Neurology* 329(4):512–518.

Snodgrass. 1981. *Oral History and Delinquency: The Rhetoric of Criminology.* Chicago: University of Chicago Press.

Spock, B.J. 1968. *Baby and Child Care.* 3rd. ed. rev. New York: Hawthorn.

Thelen, E. 1983. "Learning to Walk is Still an 'Old' Problem: A Reply to Zelazo (1983)." *Journal of Motor Behavior* 15:139–161.

_____. 1987. "The Role of Motor Development in Developmental Psychology: A View of the Past and an Agenda for the Future." *Contemporary Topics in Developmental Psychology,* ed. N. Eisenberg. New York: Wiley, 1987.

_____. 1989. "On the Nature of Developing Motor Systems and the Transition from Prenatal to Postnatal Life." In *Fetal Neurology,* eds. A. Hill and J.J. Volpe. New York: Raven Press.

_____. 1990. "Dynamical Systems and the Generation of Individual Differences." In *Individual Differences in Infancy,* eds. J. Columbo and J. Fagen. Hillsdale N. J.: Erlbaum.

Thelen, E. and K.E. Adolf. 1992. "Arnold L. Gesell: The Paradox of Nature and Nurture." *Developmental Psychology* 28:368–380.

Tilney, F. 1923. "Genesis of Cerebellar Functions." *Archives of Neurology and Psychiatry* 9:137–169.

_____. 1933. Letter to F.P. Keppel. (March 22). The Normal Child Development Project. Carnegie Corporation Papers, Rare Books and Manuscripts, Butler Library, Columbia University, New York.

_____. 1934. Letter to R. Lester. (August 13). The Normal Child Development Study, Carnegie Corporation Papers, Rare Books and Manuscripts, Butler Library, Columbia University, New York.

Tilney, F. 1968. *Master of Destiny.* New York: Hoeber, 1929 (reprinted New York: Doubleday).

Tilney, F. and L.Casamajor. 1924. "Myelinogeny as Applied to the Study of Behavior." *Archives of Neurology and Psychiatry* 12:1–66.

Tilney, F. and L. Kubie. 1931. "Behavior in its Relation to the Development of the Brain." *Bulletin of the Neurological Institute of New York* 1:226–213.

Touwen, B.C.L. 1971. "A Study on the Development of Some Motor Phenomena in Infancy." *Developmental Medicine and Child Neurology* 13:435–446.

_____. 1976. *Clinical Developments in Medicine: Neurological Development in Infancy. Vol. 58.* London: Heinemann.

_____. 1978. "Variability and Stereotypy in Normal and Deviant Development." In *Clinical Developments in Medicine: Care of the Handicapped Child,* ed. J. Apley. Oxford: Blackwell.

_____. 1994. "How Normal is Variable and How Variable is Normal." *Early Human Development* forthcoming.

Yerkes, R. 1935. *Sixth Annual Report: Laboratories of Comparative Psychobiology, Yale University* (October 1). Robert M. Yerkes Papers, University of Florida, Gainsville, Florida.

Zelazo, P. 1972a. Letter to M. McGraw (May 25) Myrtle McGraw Papers, Millbank Memorial Library, Teachers College, Columbia University, New York.

Zelazo, P., N.A. Zelazo, and S. Kolb. 1972b. "'Walking' in the Newborn" *Science* 176:314–315.

_____. P.R. 1983. "The Development of Walking: New Findings on Old Assumptions." *Journal of Motor Behavior* 2:99-137.

Zelazo, N.A., P.R. Zelazo, K.M. Cohen, and P.D. Zelazo 1993. "Specificity of Practice Effects on Elementary Neuromotor Patterns." *Developmental Psychology* 29:686–691.

PART ONE

Insights and Blunders

Introduction:
My First Science Teacher

GERARD PIEL

Myrtle McGraw was my first science teacher. In high school, I had a biology teacher and a physics teacher. They stifled what interest I brought to their subjects. My experience was not and is not today uncommon in American education. Those teachers taught their subjects from books and as closed books; biology and physics were of interest for their utility and essential to the training of a physician or electrical engineer. Already committed to a higher ambition, I managed to avoid any such further distraction at college. My A.B. in history was a certificate of illiteracy in science.

That qualification came to the attention of my managing editor at *Life* magazine, where I had begun my apprenticeship. To my dismay, he overrode my ambition to be a war correspondent (my studies had got me ready for the Second World War) and made me science editor.

My very first story was the pictorial surprise in the 26 February 1940 issue of *Life*. Engaging photographs of naked babies in a laboratory at Babies Hospital in New York City, learning to turn over, sit up, stand and walk at crucial stages in the development of their motor capacity and, then, amazing photographs of babies crawling up steep inclines and arranging pedestals to bring a lure at ceiling height in reach—all this got the science department 10 pages in that issue.

Myrtle McGraw had become my science teacher. For the first time, I saw a live scientist doing science. Science was not a closed book. Science was—as Ernest Rutherford (later, Lord Rutherford) said to Winston Churchill (then, First Lord of the Admiralty) during the First World War—"More important than your damned war!"

I would ask: "Was that from instinct?" McGraw would explain how the "closure" word halts inquiry. Evidence from observation or experiment when feasible (a narrow restriction in the case of human subjects) is the first object of the question. The asking of the question, the framing of the experiment, the evaluation of the evidence, I could see from watching McGraw at work, is a lonely and highly personal enterprise. It must proceed against hazard from what is in the books, on the one hand, and from wishful thinking, on the other. A negative answer is as good as a positive one. The reward is secure evidence that enlarges the question or proposes a new one.

I learned that the ready dichotomies—instinct *versus* conditioning, nature *versus* nurture, maturation *versus* learning—propose dumb questions. They do not set perspective for new observation nor excite productive experiment. Nothing in the Gesell model of infant development suggested that toddlers might be taught to rollerskate before they could walk. Watson had no notion that newborn infants might display a swimming reflex if immersed facedown in water.

At high voltage, Myrtle McGraw exhibited the qualities of personality I was to have the privilege of knowing in other scientists in years to come. She was free, intellectually liberated, in command of her discipline. To my wonder, she was a creative artist; yet her medium was objective knowledge. The energy and drive, tough-mindedness and tenacity, and the delight she brought to her work astonished me and made life seem a highly important experience.

These qualities I could see serving McGraw well as I learned, from her experience, that science is a public as well as intensely personal enterprise. It took some doing to get a photographer past the powers of the Columbia University College of Physicians and Surgeons into her laboratory. She was held in disdain by the starchy house officers at Babies Hospital for the publicity that had attended her work long before I approached them. The promise that photographs and copy would be submitted for review was a humiliation I had to persuade my managing editor to accept.

The photographs, by Hansel Meith, were beautiful, as the samples herewith suggest; we sketched them and made layouts in consultation with McGraw before Hansel took them. They told the story of what the study was about. My copy was institutional and insipid: here was a center for the study of child development. Nothing in the copy suggested the ferment of ideas and the philosophical controversy engaged in the work. McGraw was identified in the caption of one picture.

When McGraw lost her grants and had to terminate her work, I had my first lesson in the political economy of science in our country. Fundamental research, as I now understand, has no warrant of its own. The nearest to purity of motive for support of science is to be seen in the support of astronomy, tainted only by Wordsworthian intimations of immortality. McGraw could find no honest interest in the study of abnormal development, with therapy the ulterior motive, for which grants were now available from the foundation that had sponsored her. That she was a woman did not argue the

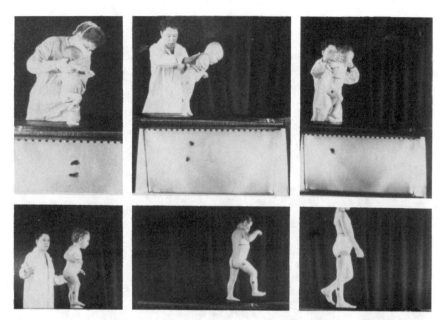

McGraw assisting a growing infant to learn how to walk at Babies Hospital, Columbia University, New York. A special podium was constructed with a rubber mat and illuminated reflecting glass sides, enabling her associates to record on film changes in stance, gait, and balance involved in the attainment of erect locomotion. (Photo courtesy of Hansl Meith and Life Magazine.)

importance of her work against the motivation and the diminution of support inevitable in the approach of war.

My tutorial with McGraw reconciled me to my assignment and started me on the invention of the magazine that became *Scientific American*. Essential in the model is the collaboration of scientists and editor she so generously engaged in with me. As the "science" budget of our country grew over this past half century, the magazine has had to reflect and report on many more episodes of diversion of support from fundamental research such as the one that terminated McGraw's work.

The diversion too often, also, reflects pollution of the thinking process in the community of science against which Myrtle McGraw so resolutely stood. She could never have been fashionable, never as celebrated, for example, as the fabulists in psychology who tout the gene for intelligence or, alternatively, for aggression and make mischief in the educational and criminal justice systems.

In all irony, the price of the neglect of pure science most often finds expression in estimates of the cost of delay or loss in practical consequences. It is

An infant reaching for a lure after successfully arranging boxes to get to the top of the pedestal, Babies Hospital, Columbia University, New York, 1940. (Photo courtesy of Hansl Meith and Life Magazine.)

impossible, of course, to calculate the infinity of loss to understanding, to human identity, to the savor of life that attends the termination or frustration of work on really important questions, interesting in their own right— as the questions asked by Myrtle McGraw remain to this day.

The company gathered in this volume to celebrate the style of scientific inquiry demonstrated by Myrtle McGraw in the study of human development may find reinforcement in the parallel life and work of T.C. Schneirla in comparative psychology. At the American Museum of Natural History, from 1943 until his death in 1968, Schneirla was curator of animal behavior (Piel 1969).

His animal was the army ant. This social insect presents, in Schneirla's (1940) words, "the most complex instance of organized mass behavior occurring regularly outside the home site in any insect or, for that matter, in any infrahuman animal." Yet, if it can be said that the discipline of comparative psychology has approached complete elucidation within its own terms of the behavioral repertory of any animal, this surely must be said of Schneirla's work on the army ants of Central America. The questions that remain from his work, that were indeed posed by him, challenge other disciplines: neurophysiology and the biochemistry of pheromones, to begin with.

The example of Schneirla's work on the army ant, stands as an enduring contribution to the intellectual apparatus of comparative psychology. On occasion, Schneirla tried to fortify the apparatus by precept, as well. He urged his colleagues to see that the terms "heredity" and "environment" overburden the simple, preliminary classification of influences in development or behavior as being internal or external to the animal. In their stead, he proposed "maturation," defined as "the contribution to development from growth and tissue differentiation, together with their organic and functional trace effects surviving from earlier development," and "experience" defined as "the contributions to development of the effects of stimulation from all available sources (external and internal) including their functional trace effect surviving from earlier development" (Schneirla 1966). External and internal influences, both, become incorporated in progressive development; the outcome of prior development is the ground of further development. In consequence, "The developmental contributions of the two complexes, maturation and experience, must be viewed as fused (i. e., as inseparably coalesced) at all stages in the ontogenesis of any organism" (Schneirla 1966).

Schneirla, like McGraw, urged the ontogeny of behavior as the proper study of comparative psychology. Such study, he argued, sets the only secure foundation for "comparing the respectively different adaptive patterns attained by animals and for better understanding of the animal series inclusively" (Schneirla 1966).

The gap "between genes and behavior," Schneirla (1966) said, is not to be bridged; it is the territory to be explored. He had in mind always that life is history. "Processes of behavioral organization and motivation cannot be dated from any one stage, including birth, as each stage of ontogeny constitutes the animal's nature at that juncture and is essential for the changing and expanding accomplishments of succeeding phases" (Schneirla 1966). The development, in its natural habitat, of an animal's entire behavioral repertory was, therefore, the necessary and proper study of comparative psychology.

References

Piel, G. 1940. "Human Infants Make Good Laboratory Subjects in Child Development Study."*Life Magazine* (February 26):51–56

Piel, G. 1969. "The Comparative Psychology of T.C. Schneirla." In *Development and Evolution of Behavior : Essays in Memory of T.C. Schneirla,* eds. L.R. Aronson, E.Tobach, D.S. Lehrman, and J.S. Rosenblatt. San Francisco: W.H. Freeman.

Schneirla, T.C. 1940. "Further Studies on the Army-Ant Behavior Pattern." *Journal of Comparative Psychology* 29: 3

_____. 1966. "Behavioral Development and Comparative Psychology." *Quarterly Review of Biology* 41:3

1

Perspectives of Infancy and Early Childhood

INTRODUCTION

When Dr. [Ronald] Oppenheim first invited me to speak to you I told him I would prefer to discuss the impact of experimental and other concepts and theories on the changing patterns of child rearing during this century. There were several reasons why I chose this subject. One is that I have been removed from infant research for many years and am not qualified to make critical appraisal of recent investigations. I am sure you are better informed on that score than I. On the other hand, in varying capacities, I have been involved with the subject of infancy and early childhood since the mid 1920s. First as a beginning graduate student and research assistant at the newly established Institute of Child Development at Teachers Columbia, later as a suburban mother and wife, and the last almost twenty years as a teacher of undergraduate students. I might say I learned as much or more in these last two roles as I did in taking graduate courses, and it was during these two roles that I became aroused about the subject matter which I shall discuss with you this afternoon.

If I draw heavily on my personal experience in presenting anecdotal material, I beg your forbearance. In substance and principle I do not believe it will vary appreciably from the points of view of my contemporaries who have lived through these rapidly changing decades—theoretically and practically, notably in developmental psychology. When I received a copy of Dr. Oppenheim's manuscript on embryology, I was relieved. He has cited so many of the authors whom I knew by their works or personally who had a great influence upon my thinking about infancy and early childhood. I hope my anecdotal presentation will be accepted as coloring to the many reviews

Lecture given at the Department of Mental Health, Raleigh, North Carolina, 1973. Included here by permission of Mitzi Wertheim.

of the precise literature in this area, of which Dr. Oppenheim's is most commendable. If at times I appear to you somewhat flippant about some past ideas and theories, it is because I prefer to be provocative rather than informative. Some of the things I say will not appear in textbooks or biographies of pioneering psychologists. Some of you may think they came out of Ripley's "Believe It or Not" but I assure you they were real at the time.

GRADUATE STUDENT INDOCTRINATION

Let me mention a couple of ideas and attitudes with which those of us who were graduate students during the 1920s were thoroughly indoctrinated. First and foremost we were told that we must learn to be *scientific*. Anything that smacked of "arm chair philosophy" was anathema to our professors. To be objective meant that one must avoid all temptations to inject their subjective or personal thoughts into their investigative procedures or interpretations. Presumably one must devise ways of collecting quantitative data which could be statistically analyzed, and then let the "data speak for itself." As is well known when it comes to behavioral research, psychology has dominated the field, especially during the early decades. It is also well known that pioneer psychologists modelled their scientific concepts and procedures on physics, Newtonian physics, which was considered one of the most scientific of all subjects. So we were taught to classify, categorize, and define terms precisely. Therefore, there was a tendency to avoid the use of common words or colloquialisms and coin new ones for the discipline. As a result a large vocabulary has accrued and has come to be known as "psychological jargon." Surely there are those among you teaching undergraduates who have experienced some students who can handle the jargon and definitions effectively, but their understanding of the real phenomenon is distinctly shallow. It seems as if having learned the verbiage, the definitions, the students experience what in Gestalt terms we call "a Closure" (jargon) and never move on to depth "cognition" (more jargon) of reality. I will have more to say about this later.

DISCIPLINE DOMAINS

While academic psychology was struggling to define and establish its conceptual territory, other academic disciplines, like sociology and education, were also on the make. Controversies ensued. I recall from my undergraduate days hearing professors in heated arguments whether or not a particular subject matter belonged in the course curriculum of one department or another. At times their arousal was so emotional I wondered if it would end up in fist fights. Usually they finally compromised on some territorial definition such as "psychology is the study of the individual" and sociology is the "study of group behavior." Such definitions created further questioning: where are the boundaries of an individual? Certainly not just the skin. I mention these conceptual confrontations and definitions merely to illustrate that our structured, traditional, compartmentalized college curricula are products of such conceptions and that they do not reflect demarcations in reality. In the old days a common result was that both departments would provide courses covering essentially the same content but adopting their own discipline-specific nomenclature. Indeed, the professors of the various disciplines became so addicted to their own jargon that they could hardly converse with each other about the same phenomena.

The outcome of this curriculum compartmentalization has been troublesome. A few years ago we saw our students protesting and shouting "irrelevant." No wonder. Now even among the traditionalist there is wide talk about interdisciplinary courses, and the college catalogs reflect attempts to reorganize course contents and to break through some of the theoretical and conceptual rigidities that result from the deeply embedded tendency to classify, categorize, polarize, or dichotomize behavioral phenomena.

DICHOTOMIES

The tendency to classify and dichotomize did not arise with the advent of the biological and behavioral sciences. We inherited it from the Greek philosophers—they polarized the "rational" and the sensual as if they were separate and distinct entities. They gave high value to the rational. That value has been dominant for the most part in Western cultures and research down through the centuries. Perhaps that is one explanation for the long delay in the study of infancy since it was claimed that the infant and young child were incapable of reasoning. But a new day is dawning. All you have to do today is read the papers, professional, and semi-professional journals to know that a wide-spread emphasis upon the sensual, feelings, and personal sensitivity is with us. People of all ages, but especially young people are joining sensitivity groups and communes. Many of the professional groups are analyzing the differences between the Western and Oriental philosophies and psychologies. It is a matter of conjecture whether these new trends will enable future researchers of infant development to break away from the limitations of dichotomous concepts. I hope so.

The most powerful dichotomy affecting our point of view about human behavior, even infant development, came in the wake of Darwin's theory of evolution—heredity versus environment. The urge to classify, coupled with the determination of the basic or smallest unit of an area of investigation, set the earliest psychologists to classifying activities in terms of: reflexes, instincts, and acquired traits. Reflexes and instincts were considered hereditary traits—fixed, immutable, impervious to environmental influences. But as the listing of instincts grew longer and longer the whole concept became cumbersome and disputable. By the time my generation of graduate students came along instinct was looked upon by some of our professors as a dirty word; we were advised to steer clear of it. Nonetheless the dichotomy of heredity-environment lingered on and resulted in the publication of volume upon volume concerned with the nature-nurture issue. What could and what could not be altered by environmental intervention was the great issue.

ARISTOTELIAN [PHILOSOPHY] AND CONFLUENCE

Next I would like to mention not a dichotomy but a kind of coming together of an Aristotelian philosophy and a Darwinian theory as the confluence affected not only early general psychology but developmental psychology. The Greek philosophers, notably Aristotle, advocated that the laws of nature were fixed and immutable. The business of the scientists was to discover those laws. You know, like gravitation, or the laws of falling bodies. (How often in real life do we encounter bodies falling through a vacuum!!). Darwin had articulated that through the process of evolution man was kindred to all other species who were also a part of nature. By the conjoining of these two points of view it was readily assumed that any principle derived from animal studies would apply equally well with humans. It is also understandable why so many of the pioneering psychologists devoted so much attention to "learning." The measuring criteria were the number of successes and failures at a given task. I am not sure about this, but my recollection is that many of the early animal experimentalists did not bother to record the age of the rats although they would report that they were litter mates. In any event, I well remember that my vintage of graduate students were certainly given the general impression that generalizations such as Thorndike's Laws of Learning represented universal verities. The focus was upon achievement, end results, more than the process by which learning took place. It is not necessary to remind you that this point of view has had considerable impact upon classroom procedures throughout many decades.

HUMAN SUBJECTS AND STANDARDIZED TESTS

Within short order school children and college students became suitable subjects for experimentation, largely because they were available in large groups and quantities of data could be accumulated with comparative ease. By that time the standardized tests (expansion of the Binet tests) had become an acceptable means against which to appraise performance in different areas. In doing so it enhanced the significance of chronological age *per se,* without which we would never have arrived at that ubiquitous symbol—the I.Q. We still live with it, we still use it as a measuring rod. Of course the standardized testing did not create the tendency to judge behavior according to chronological age but helped to turn it into a cultural obsession on the part of parents and others responsible for the upbringing of children. And along with other forces, industrialization, and public schools, it helped us structure a social system which organized human beings in terms of chronological age from pre-school to senility.

INFANCY

Beyond the familiar diaries of the nineteenth century, John Watson might be considered to be the first to subject the newborn infant to experimental procedures. But Watson's primary aim was not to determine the nature of infancy. Operating within the framework of the heredity-environment dichotomy, he wanted to ascertain the basic innate traits out of which complex behavior could be constructed. The assumption was that whatever traits were manifest in the newborn were hereditary, all else were acquired. Pavlov's conditioning theory came to his rescue—and in no time Watson became the epitome, the most vocative environmentalist. The American people liked to think of themselves as doers, so the very idea that they could develop the growing child according to their wishes by manipulating the environment was highly palatable to them. Behaviorism flourished. I shall have more to say about Watson's influence upon child-rearing patterns a little later.

It was not Watson but Lawrence K. Frank who triggered the upsurge of interest in the infant and young child during the early decades of this century. Frank was not a psychologist, a sociologist or a physiologist. He was administrator for the General Education Board of the Rockefeller Foundation. He set about funding many child development centers on many university campuses. Of course these centers attracted personnel. Many of the new experimentally oriented psychologists, having observed the effectiveness of the I.Q. with school children, set themselves the task of designing tests and scales or using such devices for the study of infants and pre-school children. Gesell's development scale for the assessment of infant development was published and available by the mid-twenties. It was then and still is the most widely used test both experimentally and clinically for the measurement of infant achievements. Like the Binet, performance of infants, in this instance, is appraised or standardized on the basis of chronological age, in months, not years. Such tests served to enhance parental anxiety if the infant did not measure up to the Gesell standards. Apparently Gesell and others found some correlation between the D.Q. (Developmental Quotient) obtained by scores on his scale and the I.Q. So far as I remember Gesell never categorically stated that infant development presaged the status of general intelligence of the child, but the implication was there and accepted by many of the users of the tests for either investigative or clinical purposes.

Gesell did something else, theoretically more significant, in my opinion. He challenged the tenets of the environmentalists, especially Watson, and in so doing he set up another dichotomy (a corollary of the heredity-environment controversy) and it came to be known as the "maturation versus learning" theory. Basically the idea was that no amount of training would advance the development or achievements of an infant until the appropriate cells of the nervous system had ripened or matured.

Watson had long since left the academic arena and joined an advertising company. However, he continued to publish in semi-professional and popular journals especially with respect to early child development. Under the influence of Watson thinking, some now amusing infant upbringing practices had become popular among a limited group of the parental population. For example, under the conditioning theory, parents should never pick up a crying infant because the baby would learn to cry to be picked up. The baby should be fed on schedule, that is, according to the clock so he would become conditioned to the convenience of the adult household; and toilet training should be introduced as soon as the newborn infant was home from the hospital. In passing, I might mention that Watson's influence on child rearing might have been different if the cultural attitude toward divorce had been at that time more liberal. If Watson had been divorced in recent decades, he would never have felt impelled to leave Johns Hopkins. [Watson forfeited his professorship when he married a graduate student.] The culture can affect the experimenter and the pronouncements of the experimenter in turn influence the culture. Surely Gesell's maturation theory took some of the pressure off parents in so far as early training was concerned. I doubt that Gesell himself ever recognized that his maturation theory together with the implied significance of the infant tests for innate endowment introduced a kind of helpless conflict for parents. On one hand, they were led to believe early development signified superior endowment and on the other hand, they were told there was nothing the parent could do about it except wait for the nervous system to mature.

I have indulged in this prolonged preamble merely to draw a kind of backdrop of *the relations between theory and concepts as they become applied in the culture of education and the up-bringing of young children.* You will notice that I have avoided mentioning that one psychological school or body of theories which has had the most pervasive and compelling influence upon American concepts of parental management than any other, namely, psychoanalytical conceptual ideas and the clinical practice of psychoanalysis. I have done so for several reasons: I'll mention only two. First of all, I am not and have never been a clinician; secondly, in the beginning psychoanalytical theories of infancy and early childhood were derived not from investigative studies of infancy but from a process of theorizing backwards from adult aberrant behavior to infancy. Everyone knows that the two systems of thinking which have had the greatest impact upon child rearing in this country during the early decades were Behaviorism, recently called "S.R." [stimulus-response] Psychology, and Psychoanalytical theories—recently called "Psychotherapy." If time permits I would like to recount for you a few episodes, personal to be sure, which illustrate the mental conflict that can go on in a student's mind when exposed to such divergent concepts.

As a young graduate student I took a course in Tests and Measurements. We were expected to go out and test a given number of students on the Stanford-Binet and the Pinter-Patterson Performance scale. We were instructed to follow the rules of the *Manual* both for administration and scoring the tests. When testing a little boy on the Binet, I asked him, "What's the thing for you to do if it is raining when you start to school?" The boy answered, "Crunch your shoulders close together and walk up close to the buildings." In New York that struck me as a highly sensible answer, but there was nothing in the *Manual* protocol of answers which would tell me how to rate him. I was bewildered.

Toward the end of my graduate career I had an appointment as a psychology intern at the Institute for Child Guidance. This organization was psychotherapeutic, and psychoanalytical in its orientation. We psychology interns were essentially a service unit, whose job it was to turn in scores for the benefit of the psychiatrist. One day I tested a little boy age eight years. His father was a cardiologist. The boy ran through the adult level of the Binet test, and finally on the Performance Kohs block tests he missed a bit, so he brushed them aside, and exclaimed, "I'm dumb." His I.Q. on the Binet was 197—I had never tested a child with an I.Q. that high, and his scores on the performance were far beyond his age level. I took him in to Dr. David Levy, an eminent psychiatrist, because I was amazed at the boy's performance on the test. Dr. Levy, a little impatient with the speed at which we psychologists got back our reports on psychological testing, had developed a kind of verbal testing for his patients. So he said to Morton, my 197 I.Q boy, "Morton, show me your palm." Morton stood there like a dumb nut. Dr. Levy repeated that for three times or more. Finally Morton pointed the finger of his right hand to the left palm and asked, "You mean this?" He simply couldn't believe that a grown-up man was asking him such a simple question. In standardized testing or verbal questioning, how often do we fail to take into account what is in the child's mind.

The impact of psychoanalytic and psychotherapeutic concepts upon the child rearing culture was too extensive for discussion here, even if I were qualified, which I am not. It should, however, be mentioned that they shifted the emphasis from performance or achievement to personal and emotional adjustment. The goal was to bring up a happy child, free from emotional conflict. The way to do that was provide him with love and more love. We were launched on what was sometimes called the "love panacea." Currently we are back on the achievement swirl, but it is called "cognition"—not intelligence or innate endowment. You can today hardly pick up a Ph.D thesis that hasn't managed somehow to get the word cognition in the title. At the same time we are witnessing an upsurge of concern with the mystic, parapsychology, psychodelic, telepathy, meditation, hallucinations, etc.—phenomena far removed from the rationality and objectivity of the 1920s and 1930s.

McGraw's Study

In 1930 I was hired by the late Dr. Frederick Tilney [Tilney died in 1938] of the Neurological Institute to provide analysis of infant and toddler behavioral development to be correlated with his studies of the structure of the nervous system. The controversy there was over structure and function. Those of us who majored in early child development at Columbia received scant exposure to neurology. Neurology was a separate discipline and was taught at the Medical School.

"In point of time" anyone venturing to study infant behavior would at the outset search for means of objective measurements. Gesell's scales had already been published. So I went to Yale, hoping to spend a couple of weeks while his cohorts, Thompson or Halverson, would give me instruction in administering and interpreting those tests. Unintentionally, but in the long run, Gesell did me a great favor. I was shown the Yale laboratory but he gave me the brush-off on the request to stay for instruction. (An aside: Helen Thompson later told me that Gesell was offended that Tilney would send a little squirt like me to visit the High Priest of Infant Development.) So I returned to the Medical Center empty handed. I was on my own. Not knowing what to do next for months on end, I spent hours on hours in the delivery room and the laboratory just looking at babies, always asking what can they teach me? As I watched, the babies became a little disenchanted with the method of checking off behavior achievements and plotting them against chronological age. It was like taking an inventory of performance. Whereas the minute changes in the quality of the behavior fascinated me, I also became interested in the influence of one developing function upon another. With these inquiries I felt the need to have at least a couple of babies, the same babies in the laboratory every day, in order to detect the subtle signals of behavioral growth. Once babies were in the laboratory on a daily basis, I seized the opportunity to do something in addition to refining our methods of observation.

You will recall that at this time the hot dichotomy in developmental psychology was *maturation* vs. *learning*. I saw no reason whatsoever to question the maturation theory. But, naive as I was, I thought that if I took a newborn and exercised it every day in any activity of which the baby was capable I would discover that magic time, that magic age, the baby could profit from practice and learning. Great! I thought I'd be able to tell parents not to bother trying to teach the baby anything until he reached the favorable age. Also as soon as a new function appeared in the infant's behavior repertoire, I tried to find ways to make it a little more difficult or challenging for him. For example, when the baby began to crawl we provided inclines for him to go up and down. I had already reached the conclusion that during infant development one could not adequately appraise an overt performance in any experimental situation without taking into account the physiological, neurological, even anatomical changes as significant factors. So I set myself the task of reading the literature in embryology, physiology, and neurology to the extent of my ability. [George] Coghill had visited our laboratory several times, and I had become somewhat knowledgeable of his work with salamanders. I was particularly impressed with his description of the "S" movement of the spinal axis of embryonic vertebrates and with his theory of individual development. I had observed that some of the newborn infants could progress several feet across the bed by using that kind of "S" spinal activity. Watson had stated on the basis of his earlier studies that there was no evidence of a swimming reflex in the behavior of the newborn infant. It was those observations and thoughts which triggered our studies of the swimming reflex of the infant.

Having been indoctrinated with the scientific idea of objectivity and measurement, the fact that I was so dependent on direct observation of overt behavior of the infant continued to bug me. Finally, a couple of young men were added to our staff—a biochemist and mathematician and a physiological psychologist. They and I vigorously pursued the development of instruments which would yield objective data. One day the young men came into my office grinning like a cat that ate the canary. They had a platform, mounted on ball bearing rollers at each of the four corners. The corners were to be attached to a kymograph, so we would have four recordings of the baby's balancing stability in the development of walking. That, I said, might give us a measure of balancing, but it wouldn't tell us anything about the child's propulsion or stepping aptitude. The way these two forces work together is what is important. Ball bearings! ball bearings! A thought. Why don't we put him on ball-bearing roller skates, I asked. We had to have the skates custom made because no one had ever heard of putting a year-old infant on skates. We did just that, and that was the beginning of our study of the skating behavior of infants.

When I first reported on our studies illustrated with movies, to a meeting of the APA [American Psychological Association], Gesell and my other professional colleagues jumped to the conclusion that my findings negated Gesell's theory of maturation. The idea of babies, even newborns, making swimming movements and an infant of 13 months scooting about on roller skates made the newspaper headlines. In those days making the newspaper headlines was no asset to the serious investigators, especially among one's conferees. Furthermore, the total study in no way negated the maturation theory, it only added certain refinements. For example, at the same time the year-old baby was put on roller skates, he was also placed on a tricycle, with his feet strapped to the pedals. The entire culture assumed that a child would learn to tricycle before he learned to skate. But for nine months, every day, five days a week, we pushed that baby up and down the corridor on the tricycle. He knew what we wanted him to do, and he would imitate our efforts to start the vehicle. But he just couldn't get the idea that it was his own pressure of his feet on the pedals that made the vehicle move. Then one day when he was about 19 months old the dawn came. The glint in his eye, the facial expression of surprise, were wonderful to see when he recognized that the power of his own body, his feet, made the tricycle move.

Now let me tell you something about the academic, experimental pronouncement and the culture of the society at large. Neither my professional colleagues who wrote text books nor the popular journalist gave attention to my published report of the tricycling. In the minds of the public there was nothing phenomenal about a toddler learning to tricycle—so the long nine months of effort was overlooked. The swimming and skating, being novel, got all of the attention. For example, it was the facial expression, the glint in the eye, more than the leg movements which told me that new pathways in the baby's brain had begun to function. Yet, recently, when I looked over my original publication I noticed I had not mentioned the facial expression. I had, as a student, been so indoctrinated with the objectivity goal I was afraid I would be accused of being subjective or sentimental had I done so. Also for years, because of the extensive notoriety given to the swimming and skating performances, textbook publishers cited newspaper and magazine articles and not from the published report of the entire study.

As stated above both the indoctrination of students, future experimentalists and the milieu of the culture affect the interpretation of an investigation, which in turn through the media affects the thinking and attitudes of the society. One could go on ad infinitum citing anecdotal illustrations of the influence of academic theory and concepts upon society and the impact of the zeitgeist or social ambience on experimental investigations. I would like to take a few remaining minutes to throw out some opinions, personal opinions, not as dicta but as challenges for your consideration and discussion.

In the area of infancy and early child development a new day is dawning and you young researchers have a golden opportunity to clear up some of the morass that has beset us during this century.

(1) Happily, there is an emerging attitude that scientists, all scientists, should assume some responsibility for the way in which their pronouncements are adopted into the culture. This is most important in connection with recommending child rearing patterns.

(2) Our traditional academic disciplines are such by definition only, and do not reflect lines of demarcation in reality. Fortunately, there is a growing trend for interdisciplinary research and interdisciplinary instruction. Professors should be constantly aware of the conceptual indoctrination which they pass on to their students, future investigators.

(3) There are two concepts which, in my opinion, are universal representing realities as we experience them. One is the *process* of growth and the other is communication theories. I believe if we could somehow bring together a conjoining of these two concepts, articulate them well, they could provide better approaches for research and better guidelines for parents, the caretakers of infants and young children. Both of these concepts represent something constantly in flux, never static. Years ago I became disturbed that developmental psychologists had modelled their procedures after general psychology, who in turn had modelled their procedures after physics. As a result we began to compartmentalize the child: Look into the early textbooks. You will find chapters on mental, emotional, social, physical development, individual differences, etc. etc. If at the outset we had modelled our experimental procedures after embryology or biology we might have been able to formulate theories better suited to deal with constantly changing phenomena, and with the interrelations of one system to another. When I first began to read about computers in the mid-fifties there was considerable discussion about the potential analog computer. That was exciting because it seemed to offer the means of getting a running account of communications among multi-systems. Now we don't hear much about the analog computer. What a pity!

(4) I hope your generation may be able to free us from some of the culturally ingrained dichotomies which clog our own mental processes.

(5) Future investigators would be wise, in my opinion, to use common words whenever possible rather than increasing the bulk of professional jargon. My observation is that both students and parents get so involved with the professional jargon that they don't really comprehend the phenomena it presumably represents. Scientists, all scientists, but especially behavioral scientists, have a tendency to introduce new nomenclature because they want to be sure of the clarity and precision of their concepts. But in this day of mass media, once those concepts have been introduced into society, they take on many different meanings, a coloration, more embarrassing than the

specificity the original scientist intended. In other words, they take on the disadvantages of common words. (Illustrations: "conditioning" and "feedback.")

(6) Behavioral scientists, including developmental psychologists, should try to establish a working partnership with other institutions of the society—the mass media, business, architects, and government. A metamorphosis seems to be taking place in most of our established institutions. The question is: Could we through a working partnership steer and direct the course in which they go? For example, one of the greatest needs today is to bring about an intermingling of the generations from birth to senility. How can we do it?

(7) Professionals and educators should struggle to design ways of cultivating "intuitive sensitivity" of the learner at the same time that one is learning about infancy and early childhood. Intuitive sensitivity cannot be gained merely by reading books, listening to lectures, or attending classes.

(8) Why can't we set up a computer system so that investigators could renounce their expressed ideas which they have later conceded to be erroneous rather than having them survive in the references or bibliographies.

❧ 2 ❧

Professional and Personal Blunders in Child Development

At the dawn of the century many new disciplines, psychology among them, were incorporated into the curriculum of prestigious universities. Due to the promotional activities of L. K. Frank of the General Education Board of the Rockefeller Foundation, centers for the particular study of child development were established at various universities throughout the country. Columbia University was a mecca for graduate students eager to specialize in these new disciplines. I was one among them.

Having completed the requirements for the Ph.D and also having spent a year as a psychology intern at the Institute of Child Guidance, I was moderately familiar with the significant studies and the controversies about dichotomies going on at the time, such as nature-nurture and maturation-learning. Standardized testing techniques for evaluating infancy and early childhood development were in vogue.

In 1930, I was assigned by Dr. Frederick Tilney of the Neurological Institute to undertake a study of infancy, beginning at birth of the children and following them through the early years, to provide him with a behavioral protocol to accompany his investigations of the cellular development of the brain and nervous system of the human. At the time there was no well-staffed center for the study of infancy connected with the Neurological Institute. By a streak of good luck, space for a laboratory was provided at Babies Hospital to accommodate the follow-up study at regular intervals of the groups of infants of the original assignment.

The nature-nurture issue, otherwise labeled heredity vs. environment, was a prominent controversy among more established investigators, as revealed in the literature. The assumption was that any behavioral traits present at

Reprinted from *The Psychological Record*, 1985, 35:165–170 by permission of Kenyon College and Mitzi Wertheim.

birth could be ascribed to heredity, and traits that developed after birth should be attributed to environment.

It was thought that if we could have one or two infants brought to the laboratory for the day five days a week, and provided one of them with challenge and exercise for any new trait appearing, we might be enabled to make a contribution to that particular controversy. We selected Johnny and Jimmy for that particular study, in addition to our major assignment. Johnny was selected as the experimental subject for the special study. The outcome was published in *Growth: A Study of Johnny and Jimmy.*

All was going well for a year, without a hitch, until Johnny began to stand for a brief moment and manifest the urge to try to lift a foot. From the outset our problem was how to evaluate Johnny's improvement in equilibrium and stepping, and how to challenge and practice on those interacting aspects of neuromotor development. It was in this connection that I performed my most troublesome fumble, or blunder, in designing a challenge and practice for Johnny.

Before presenting details of this blunder I invite readers, especially the young, to indulge in a little speculation or imagination:

> Assuming you have conducted research on infant and early childhood development for several years, and that when you have all your findings ready for publication, you discover that the outcome of your study is contrary to what you had anticipated. Not only contrary but also utterly different from the expectations of the general public. You are puzzled, confused as to how to explain the divergence in some aspects of your findings and general expectations.

That is the situation in which I, as a young researcher, found myself some 50 years ago. The bewilderment lasted many years. It is the final resolution of that bewilderment that prompted me to write this essay. Any experimenter is likely to have made mistaken judgments or blunders. If this confession of some of mine can be of comfort to young researchers and help them recognize their own mistakes, then I am rewarded.

BLUNDER #1 SKATING AND TRICYCLING

As revealed in *Growth: A Study of Johnny and Jimmy,* in every activity throughout the first year of Johnny's special exercise or practice we took the lead from the infant's behavior, the signals of behavioral change, to select appropriate challenges and practice of the emerging new activity. For example, when he first began to make creeping movements we provided inclines of different steepness which he could climb up and down. But when it came to standing and trying to step we were, at first, puzzled as to how to prepare the favorable challenge for equilibrium and stepping. The idea of ball-bearing roller skates was suggested. We all laughed, thinking it ridiculous. But we did just that. We had a pair of ball-bearing skates custom-made and attached to the soles of a pair of shoes that fitted him.

I don't recall just how the thought of a tricycle first arose, but we were familiar with Raven's report of teaching a chimpanzee to ride a tricycle. So we bought a tricycle suitable for Johnny's size and fastened the soles of his shoes to the pedals. Johnny was first exposed to the roller skates and the tricycle on the same day, just about three days before his first birthday.

Of course we never had the slightest idea of getting that baby to learn to roller skate; we just wanted to see how his growing neuromotor system adjusted to an unfamiliar situation. We certainly took it for granted that he would learn to manipulate the tricycle, with his feet tied to the pedals, long before he would be able to perform well on skates. We were in for inexplicable surprises. Within 2 months, at the age of 14 months, Johnny was performing amazingly well on skates, and not only on flat surfaces. He was able to coast down a slightly inclined hallway—and did so with glee. During that same 2 months on the tricycle he knew what he wanted to do, and by shaking his body forward and backward he tried to get the vehicle to move. If we attempted to give him the sensation of the vehicle moving by giving the rear wheel a little push, he would take his hand and try to start it but never with enough force.

I knew that this baby would learn to ride that tricycle some day and I was determined to find out when and how. So for 7 months, twice a day, 5 days a week, we hitched his feet to the pedals and tried to lure him forward. But during those 7 months no appreciable improvement was manifest.

Then one day when he was 19 months of age, I stepped into the laboratory after his feet had been tied to the pedals and Johnny looked up at me as if he had discovered a law of the universe. He had just realized that it was the pressure of his leg on the pedal that started the vehicle to move. There is no way to describe the emotional communication and delight shared by me and the baby at that moment! The recollection of it is just as vivid today as it was on its occurrence. Within 10 days Johnny had mastered the tricycle, but he never manifested the delight of the accomplishment as much as he did when he skated when only 14 months old.

For me the problem was augmented. For years I searched my mind for any explanation for the comparison of that baby's unexpected performances on skates and tricycle. It was generally accepted in psychological investigations that if a scientific design had proved its worth in a study of animal behavior, especially of a primate, it was reasonable to use the same technique in a study of humans. My former professors and colleagues were [so] impressed when seeing the skating films of Johnny that they tended for the most part to

overlook the report of his tricycling. Then one day when I was not consciously searching for the explanation, heuristically I realized that my blunder was in adopting the Raven technique in introducing the baby to the tricycle. Tying his feet to the pedals confronted that year-old infant with a
situation in which he would have had to perform completely or not at all a
very complex action. In his confusion all he could do was the random shaking of his body. In all the other activities to which he had been exposed we
designed the set-up according to developmental signals detected from the baby's actions. When this was done he achieved astonishing results, as demonstrated by the skating. So it was I, the experimenter who was responsible for
those perplexing results. I might add that even the toy manufacturers of
those days were more perceptive. In their designs they took into account the
infant's state of neuromotor development by designing a "kiddie-car" for
sale for infants before suggesting a tricycle! Now, in retrospect, I am quite
confident that if we had not tied that baby's feet to the pedals he would have,
by virtue of his own investigative urges, placed his free feet from time to time
on the pedals and by those actions discovered earlier that it was leg-pushing
on them that brought about movement of the vehicle. So I, the experimenter,
blundered, not the baby.

Blunder #2 Age in Days

The report of this second blunder I lay somewhat at the directive of my scientifically oriented graduate professors. We were instructed that when designing an experiment we should start with the lowest common denominator of
any aspect of the phenomenon being investigated. Chronological age was an
aspect of child development and the lowest common denominator would be
age in days.

But my real motivating force for recording the age in days was social. I had
become skeptical of the influence of standardized testing during infancy and
critical of the impact of the advice of the "experts" and the directives of
"how to do it" books upon parents. The impact was such that parents
wished to see their infant not in terms of individual development but how
that child compared with the hypothetical average for chronological age. I
thought that my recording age in days might make a dent in that compulsive
social pattern. In that respect I not only blundered, I failed.

I discovered my blunder when I began teaching undergraduate students.
When they read my early publications they would pause in their reading of a
sentence or a paragraph in order to calculate the "age in days" into weeks,
months, or years! So age in days served as an interference or interruption in
their comprehension of the thoughts presented.

BLUNDER #3 CHAPTER 7 OF
CARMICHAEL'S FIRST EDITION

By the mid-thirties investigations and publications on the subject of child development were flourishing in the centers and universities throughout the country. Theoretical controversies were common and exciting. *Growth: A Study of Johnny and Jimmy* was published in 1935. Work at our laboratory at Babies Hospital was thriving and running smoothly with the aid of a competent staff.

During that time of a favorable atmosphere for intellectual inquiry I was invited by Leonard Carmichael (1946) to contribute a chapter to the *Manual of Child Psychology* that he was organizing. I gladly accepted his invitation and chose "Maturation of Behavior" as the topic for my chapter. At that time one of the most acrimonious disputes emanated between Gesell and the behaviorists, especially John B. Watson, and it was formulated in the dichotomy *maturation versus learning*. So the subject for my chapter was timely.

Never before nor since have I worked so diligently or covered the literature so extensively—literature not only in child psychology but also in embryology, genetics, and other allied subjects. When it was finished I was proud of it. (Rereading it recently in preparation for this essay, I am still proud of it, recollecting the spirit of the times in which it was compiled).

The manuscript was turned over to Dr. Carmichael in 1940–41. National and international tensions were high and World War II was imminent. Our laboratory was being terminated and I was winding up my activities at Presbyterian Hospital. My family moved to the suburbs. During those eventful years the thought of the whereabouts of my manuscript never entered by mind.

However, in 1946 the first edition of Carmichael's manual was published. Shortly afterwards I received a request from him to survey the recent literature in order to bring the material up to date for the preparation of the second edition. I thought, heavens to Betsy! The first edition had just been published—completely forgetting that my chapter had been completed 5 years earlier. Anyway I was in no position to do an extensive study of the literature at that time and I was unwilling to turn in a make-shift addendum to the chapter on which I had worked so diligently. So I wrote to Carmichael stating that the literature in the original chapter was sufficient to conclude that the disastrous dichotomy of maturation and learning was untenable and that perhaps it should not be rearoused. That was when I committed the most colossal publication blunder of my career. It was a blunder because it did not dampen the controversy of maturation versus learning, it merely aroused curiosity among colleagues as to what prompted the deletion of Chapter 7 from the second edition of Carmichael's Manual. This is illustrated by the following footnote taken from Gilbert Gottlieb's (1976) article "Conception of Prenatal Development" in *Behavioral Embryology*:

4. Thirty years ago, the developmental psychologist McGraw (1946) recognized that the structure-function relationship could very well be bidirectional in her study of the early motor development of the human infant. She stated: 'It seems fairly evident that certain structural changes take place prior to the onset of overt function; it seems equally evident that cessation of neurostructural development does not coincide with the onset of function. There is every reason to believe that when conditions are favorable function makes some contribution to further advancement in structural development of the nervous system. An influential factor in determining the structural development of one component may be the functioning of other structures which are interrelated' (McGraw 1946:363). As Crowell (1967) has pointed out, although McGraw did not work out a systematic theoretical position, her notions did moderate the more extreme interpretation of Gesell's (1946) 'maturational hypothesis' and thus were a step toward an appropriate resolution of the nature-nurture problem. In light of the importance of McGraw's conception and its probable correctness, it is interesting to note that McGraw's (1946) chapter was deleted from the second edition of Carmichael's Manual of Child Psychology (1954), while Gesell's chapter (1946;1954) was retained.

BLUNDER #4 "CRITICAL PERIODS"

This item is not so much a blunder as a warning to present day investigators. Currently there is a recognizable, tangible tendency in academic circles toward a synthesis of cross-disciplinary thinking and verbal usage. In my early publications I made frequent use of the term "critical periods." I had probably picked it up from my extensive readings in embryology. Being impressed with the disarray that is manifest as a newborn's reflexes begin to decline when the signals of new behavioral traits emerge, and with the fact that those signals indicate a readiness of challenge or practice, I referred to that type of transition as "critical periods" for advancing behavior. But in the course of events I discovered that many writers were assuming that if opportunity was not provided at that time the behavior would never be achieved. It was certainly not my intention to imply that if a particular trait was not encouraged at the "critical" time it would never be achieved. I did write an article admitting the misapplication of the term and suggesting instead an "opportune time" as an appropriate label for the nurturing of a particular trait. When adopting a catchy term or label from one discipline to another it would be well to make sure it doesn't have implications that would not be applicable to the phenomenon under consideration.

Blunder #5 Theories

Over the years colleagues have commented that I never attempted to formulate a McGraw Theory of early behavior development, as illustrated in the footnote quoted from Gottlieb's article. An explanation for my failure to do so is quite simple. At the time I was not qualified with an understanding of the artistry and techniques of theory formation, nor did I have the mathematical skill to do so. Perhaps therein lies the blunder. My concept was of multisystems developmental processes emerging and advancing at different times and different rates, but finally interacting, integrating, and synthesizing for the creation of new performances or traits. I preferred to present my findings as observed and to allow future researchers to make use of them as they saw fit. Had I attempted to formulate a theory for such a complex of processes and then decorated it with a catchy acronym, the chances are that it would soon have been challenged by some current or future investigator and then we would have another troublesome dichotomy to deal with. Call it a blunder or a deficiency of the experimenter if you choose. Perhaps some future or present day investigator can formulate a comprehensive theory of development that can withstand dichotomies. The subtle, complex processes deserve a reliable, workable theory.

Blunder #6 SRCD Membership

During the 1940s with the nation in the midst of war and having relinquished full-time research and moved to suburbia, I entered upon what I later called my "D-D Decade" (Domesticity-Diversity). Thereupon I withdrew my memberships from the many professional societies of which I was a member—APA (American Psychological Association), AAAS (American Association for the Advancement of Science), SRCD (Society for Research in Child Development) et al. I did so because I was very much occupied with other activities and also because I felt those societies should not become loaded with nonprofessionals. Furthermore, I chose not to call on my husband to pay my membership fees and the cost of attending meetings. This was my most stupid blunder!

When I entered teaching during the early 1950s I rejoined APA and some of its divisions. Although SRCD was my favorite society, I did not ask to rejoin it. In my mind it was the one professional society that represented exclusively persons engaged in research in the processes of growth and development of the human infant and young child. I did not want to be a part of contributing to SRCD's becoming as diversified as had APA. That attitude was perhaps the most senseless blunder of my professional career. I should have rejoined. The urge for exploratory activity stayed with me so much that even when teaching college students I organized a teaching laboratory for instructing them in the art of observing and reading the signals of the prelinguistic infant and young child. So my concern for exploratory investigation in child development joined with my concern for educational procedures.

If this confession can help educators and young researchers recognize the value of admitting blunder or mistaken judgment the process of growing up in this society can be improved.

PART TWO

The Maturation Controversy

Introduction:
Johnny and Jimmy and the
Maturation Controversy:
Popularization, Misunderstanding,
and Setting the Record Straight

PAUL DENNIS

Few studies conducted by psychologists have received the amount of publicity accorded McGraw's 1932 co-twin study of Johnny and Jimmy Woods. Containing the ingredients necessary to a good news story, Johnny and Jimmy became minor media celebrities from 1933 to 1942 and the McGraw experiment somewhat of a media event (Dennis 1989). However, in selecting certain stories to report the press defines which scientific issues are newsworthy and which are not; and, in their choice of words and metaphors they convey certain beliefs about how science works and what social meaning is to be attached to its results (Nelkin 1987:88). In the case of newspaper and magazine coverage of Johnny and Jimmy Woods, the press emphasized the issues of environment and heredity, the role of early experience, the merits of behaviorism, and psychology's potential to contribute to improved techniques in child-rearing. However, they failed to examine these issues in any depth, overestimated the ability of McGraw's study to resolve them, and neglected completely to reflect the fact that while psychological research often sharpens our understanding of an issue and adds to our knowledge, only rarely does it resolve in a single stroke such complex issues as those addressed by McGraw.

At the time McGraw conducted her study the attention of an entire nation was focused on two of the most important news stories of the decade, the kidnapping of the Lindbergh child in 1932 and the birth of the Dionne quintuplets in 1934. That the press would choose to publicize yet another set of babies was not surprising in light of the novel methodology employed by McGraw and the initial dramatic reports of Johnny climbing steep slopes, swimming with his face under water, and roller skating at fifteen months of age (McGraw 1935). Unlike many behavioral science studies that reduced social phenomena to statistics and held little human interest, McGraw's test of the effects of stimulation and restriction on two little boys possessed human drama capable of sustaining long-term reader interest.

In addition, her study dealt in controversy. During the 1930s thoughtful citizens, social scientists, and parents hoped that the implementation of new ideas in child-rearing might produce more socially responsible adults in the future who would be better able to resolve the increasing social problems besetting an already growing industrial nation. However, parents were confronted by differences between psychologists regarding the relative significance of environment and heredity, maturation and training and the issue of possible harmful effects resulting from the over-stimulation of infants. In particular, the debate between Arnold Gesell, who had gained a large popular audience in the 1930s and 1940s through his syndicated columns and books, and John Watson, whose popular articles also had reached large numbers of readers, posed a confusing set of directives to parents regarding the relative importance that was to be attached to training and maturation.

The results of McGraw's now classic experiment were described in various newspapers, including the *New York Times* and written up by such publications as *Newsweek, Parents Magazine, Commonweal,* and the *Literary Digest.* "'Conditioned' Infant Excels Twin Brother," wrote the *Literary Digest* 1933a:17), and "Is a Race of Supermen Really in Sight?" asked *Parents Magazine* two years later (Hansl 1935:57). Citing the initial reports of Johnny's extraordinary motor accomplishments as evidence for McGraw's study having "far-reaching consequences to child-rearing methods" (*Literary Digest* 1933b:18), the press also described what they perceived to be the effect of the early training on the twins' personalities. Journalists pointed out that Johnny was "superior mentally,"better mannered, and more "independent, resourceful and sure of himself" than Jimmy who was described as "full of pranks,"more aggressive and less well-mannered (*Literary Digest* 1933a:17). However, when McGraw (1935; 1939) reported that Johnny's dramatic motor performance began to fade after stimulation was halted, and with training Jimmy's motor skills improved and approximated those of Johnny's (Jimmy had received two and one-half-months of intensive stimulation following the removal of earlier restrictions) the initial enthusiasm over the study showed by the press quickly changed to a mood of disillusionment. Reporters now claimed that an infant's earliest training was "not retained" (*New York Times* 1934a:7) and that "the early months in a child's life did not count" (*New York Times* 1935a, n.p.). The *New York Times*

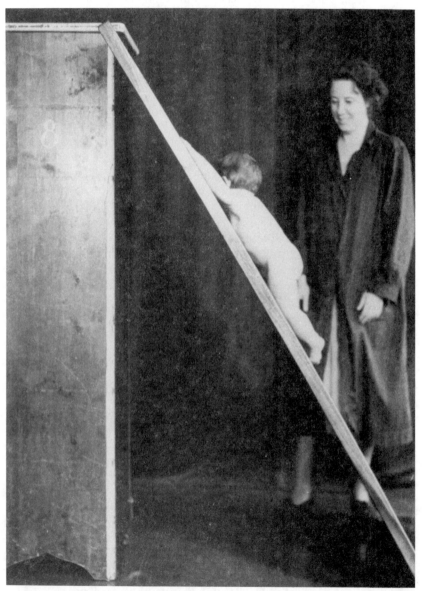

The "experimental twin," Johnny Woods, ascending the slide under McGraw's watchful eye at 22 months of age, Babies Hospital, Columbia University, New York, 1934. (Photo courtesy of Mitzi Wertheim.)

(1934b:4) asserted that while her "results will disconcert those behaviorists who had hoped that a complete regeneration of humanity might be achieved in short order by conditioning every baby properly in its first year," it would "doubtless relieve some parents who had taken their responsibilities too heavily fearing that one false conditioning, as a result of accident or carelessness, might blight their child's character and distort his emotional life forever."

In addition to conceptualizing McGraw's work as a critical test of behaviorism, the press went on to suggest that the study was a test of psychology's potential to contribute to better child-rearing practices. Journalists now contextualized the experiment as a contest between "Just Plain Jimmy, who was brought up in the same manner as any normal boy" (New York Times 1942:40; Literary Digest 1933a:17) and "Scientific Johnny," who had benefited from "university" training and the "best scientific knowledge as to child behavior" that psychology was able to offer (New York Times 1942:40; Literary Digest 1933a:17; New York Times 1937a:25). While ignoring the important long-term differences in motor processes and attitude noted by McGraw, as well as Johnny's remarkable earlier motor performance, newspaper accounts rallied to the support of Jimmy as if he were the underdog in some sort of undemocratic experiment because of the early special treatment accorded Johnny. On the occasion of one of the twins' yearly birthday trips to the circus, the press claimed that Jimmy was more natural in his reactions than was Johnny who appeared "moody," "very bored," and "conditioned" in his reactions; "if the audience clapped," wrote the New York Times (1939b:20), Johnny "looked up from his Mickey Mouse and clapped, too."

When not proclaiming the virtues of Jimmy ("Unconditioned Twin,...Makes 'Scientific' One Seem Tongue-Tied" and "Brother 'Bosses' Scientific Twin,'" headlined the New York Times 1941:16; 1938:8), journalists noted either "no salient differences" or argued that while Johnny might be "better equipped for life than his unconditioned brother" in the rarefied atmosphere of the clinic, "in their home the 'advantage' was not so apparent" (New York Times 1937b:8; Literary Digest 1933a:17). Harshly criticizing psychology's ability to contribute to better child-rearing practices, the Commonweal (1939:3) was to write of McGraw's study: "it is, not that all of this proves anything, but that it proves nothing. In other words, the clinic had better have its 'conditioner' overhauled. We very seriously doubt that it had any effect on Johnny, one way or the other. Perhaps it would be even better if the clinic threw the conditioner away altogether."

Although journalists had oversimplified, sensationalized, exaggerated, and misinterpreted the results and purpose of McGraw's experiment during the course of press coverage, there is little evidence to suggest that psychology texts attempted to correct these misunderstandings. As noted by

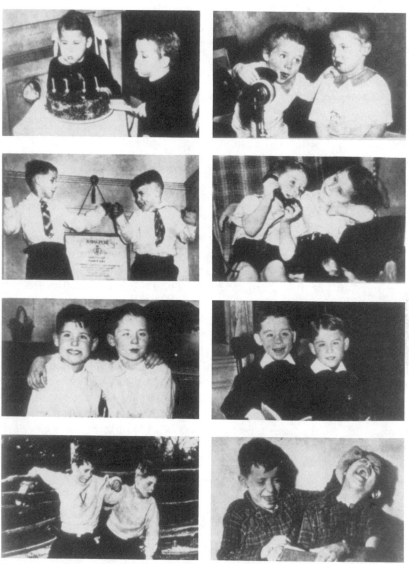

Johnny and Jimmy, ages 3 through 10. *Top row (left to right)*: Normal 4-year-old interest in a birthday cake; Jimmy *(left)* at the microphone demonstrates his amusing
line of chatter. *Second row (left to right)*: Johnny *(left)*, at 5, still handles his muscles
more smoothly; at 6 Johnny is quieter, Jimmy, the more gregarious twin. *Third row
(left to right)*: At seven, Johnny *(left)* is more imaginative, Jimmy, the happy-go-lucky
twin; Johnny *(right)* and Jimmy grin and bear another birthday "shot," their 8th.
Fourth row (left to right): Normal boys of 9, Johnny *(left)* and Jimmy are unspoiled
by public notice; Johnny *(left)* and Jimmy are healthy, active boys at 10 (McGraw,
1942:23). (Photo courtesy of Mitzi Wertheim.)

McGraw, "Many of the quotations in the textbook for years was from these magazines and newspaper reports, and not from the book that I had worked so hard to get together" (Senn 1972:17). In both cases, the main conclusion rendered by the press and psychology was that heredity and maturation, not environment and training, principally determined the course of motor development. As recent as 1993, authors of one child psychology text, Hetherington and Parke (1993:179), labeled McGraw a "maturational theorist" while a text in developmental psychology by Shaften (1993:171) cited McGraw, along with Gesell and Thompson, as investigators whose work supported the "maturational view-point." (See also Razel 1988.) [An extensive search by the editors turned up only one journal review of McGraw's book, *Growth*, by Sherbon (1938), a female physician. Despite this meager coverage, Sherbon provides a remarkably thorough, accurate, and insightful treatment of McGraw's research.]

Aside from the obvious role that reliance upon secondary sources has played in perpetuating misconceptions about McGraw's study, confusion as to how to properly interpret her results in all likelihood would have occurred even in the absence of extensive press coverage. (See also, Bergenn, Dalton, and Lipsitt 1992.) More specifically, the unforeseen failure of McGraw's study to adequately satisfy the requirements of a co-twin design (it was generally assumed afterwards that Johnny and Jimmy were fraternal rather than identical twins) rendered problematic the question as to whether differences between Johnny and Jimmy were due to either maturation or training. Because the prevailing tendency in psychology at the time was to emphasize maturation, the failure of Johnny's initial motor accomplishments to persist were interpreted as consistent with a maturational viewpoint. And, to the extent that differences in motor processes cited by McGraw were even considered, they probably were ignored for the same reason. However, less understandable was the neglect by psychologists of Johnny's extraordinary early motor accomplishments. Exceeding not only the performance of Jimmy but that of all the controls, it was clear that "training" (as used by McGraw) had made a difference despite the claim of maturational theorists that training should not make a difference. The fact that Johnny's motor performance was not maintained might have reflected either the limitations inherent to the training procedures employed by McGraw or the tendency for unpracticed skills to decline over time rather than the inherent limitations imposed by maturation.

In addition, McGraw's conceptualization of the role played by training in early motor development was conceptually challenging in its attempt to transcend the prevailing environment/heredity and training/maturation dichotomies, as well as in its disagreement with models of learning that emphasized stimulus and response, association, and conditioning. In the first instance, McGraw argued that neural growth processes exhibited plasticity and adaptability, and that the truth regarding the role of training lay somewhere between the maturational claims of Gesell and environmental emphasis of Watson (McGraw 1942). However, by acknowledging the importance of both environment and heredity, (and maturation and training), but leaving unanswered which is more important, she implied a more complex and relative solution to a press and psychologists who preferred a simple and absolute answer. Suggesting that there was a time when a child could profit from training and exercise, McGraw (1940) argued that in order to find out when this might occur one had to learn to read the child's signals and know how to arrange the child's environment so that it posed a challenging experience.

In the second instance, McGraw saw the study of motor development as demonstrating that learning entailed more than simply the gradual enlargement or mechanical recombination of specific motor skills through simple conditioning. Alternatively, she suggested that the formation of more developed and varied patterns of behaviors in motor development was a dynamic process involving both structural and functional components as well as thoughts and feelings. As a consequence, the "training" or "stimulation" provided Johnny through the use of lures, suggestion, and an alteration in context to provoke problem solving behaviors was an attempt on her part to teach him to maximize his potential while not interfering with his natural reactions (McGraw 1935).

For readers unfamiliar with McGraw's original work and ideas, the essays included in this section help clarify those findings and concepts which have been most frequently misunderstood. "Later Development of Children Specially Trained During Infancy" and "Experimental Twins" review the purpose and context for McGraw's study. The latter essay (subsequently revised and published in the New York Times Magazine in 1942) was written in order to address public misconceptions about her study. "Later Development" was published in Child Development as a follow-up study to her book Growth: A Study of Johnny and Jimmy, 1935. Both essays make clear that the purpose of her study was to examine the phases of development a growing infant goes through in order to determine whether or not the achievements of a given motor performance might be altered by daily stimulation or restriction of an activity. Although not of the opinion that environment had such overwhelming significance as claimed for it by Watson, McGraw disagreed with Gesell's assertion that there was no conclusive evidence that practice and exercise accelerated the appearance of various motor behaviors. Moreover, these two papers, as well as her 1986 essay, "The Problem of Using Secondary Sources: Elkind's Blunder" make clear that the study was not an experiment in "conditioning" nor a test of behaviorism. Finally, McGraw reiterates in these essays, and in a 1958 paper presented at a Soviet sports conference, her contention that exercise during infancy does have permanent effects and that there was a long-term difference in motor processes between Johnny and Jimmy. Although the differences reported were not apparent in measures of performance, the quality of Johnny's motor coordination was greater than that of Jimmy's at ages six and twenty-two. Likewise, McGraw (1939:12) points out in "Later Development" the importance of attitudes in facilitating and inhibiting motor behavior, by noting that Jimmy was always more concerned with "end result" while Johnny was more concerned with "modus operandi."

McGraw argued in "Experimental Twins" that if babies between six months and two years of age were given a chance to exercise their muscles they would use them "more smoothly and with better coordination later on." She also attempted to specify when the most effective time for such stimulation might be. She referred to such times as "critical periods" in "Later Development" (McGraw 1939:2). However, because of the maturational overtones implied by the widespread subsequent use of this term, McGraw's earlier usage is likely to be misunderstood. She did not intend to imply that behavior could never be later developed if stimulation did not occur during this period. In her 1940 essay, "Signals of Growth," McGraw makes clear that her use of the term critical period is to be understood within the context of an interactive relationship between structure and function and the regression and progression of behaviors. With this in mind, she emphasized to parents the importance of recognizing the "signals" in a child's behavior which might be used as a guide to knowing when and how much a child should be stimulated.

Finally, an unpublished 1986 essay written when McGraw was eighty-seven years old, "The Problem of Using Secondary Sources: Elkind's Blunder," highlights the continuing concern she had regarding the various interpretations that had been given to her study of Johnny and Jimmy. Responding to David Elkind's (1986:636) citation of her work as an example of "too much adult intervention in the self-directed learning of young children," McGraw describes the nature of the stimulation provided to Johnny in order to counter the claim that her study was an experiment in conditioning or teaching techniques and relevant to the issues of interest to Elkind. In addition (and as an example of how the use of secondary sources has plagued an understanding of her study) she addressed Elkind's report of a conversation he had with a psychiatrist (presumably William S. Langford) who had interviewed Johnny and Jimmy in 1939 (See McGraw 1939:12–16). Elkind cited Langford's opinion that Johnny was more insecure and dependent on adult approval than was Jimmy as anecdotal evidence for the potential risk to which children were put if parents intervened too much in the self-learning process.

Interestingly, the claim made by Elkind was similar in kind to various claims made almost fifty years earlier by the press regarding Jimmy's alleged better adjustment in a variety of areas. When the twins had reached the age of nine, the *New York Times* (1939a:19) wrote: "Johnny Woods Only a Little Above Average in Studies–Brother Who Had Untutored Infancy Gets Almost Perfect Grades." However, just as the press had ignored the context for the claims it made (in all likelihood based on Langford's analysis as reported in "Later Development"), so did Elkind fail to take into consideration significant differences in attention provided the twins when they were at home rather than in the laboratory. Langford (see McGraw 1939:13–16) noted that the twins's parents could not resist favoring Jimmy when home, by encouraging him to express himself and take the lead. McGraw (1939:17) observed that Johnny had difficulty adjusting to a home situation in which his skills went unacknowledged and consequently was more withdrawn. McGraw concluded that given such confounding influences, "there is no way to evaluate the extent to which the early experimental experiences operated in determining their (Johnny and Jimmy) respective personality components" (McGraw 1939:19).

References

Bergenn, V W., Dalton, T.C. and Lipsitt, L.P. 1992. "Myrtle B. McGraw: A Growth Scientist." *Developmental Psychology* 28:381–395.

Commonweal. 1939. "The Woods Twins." 30:3

Dennis, P. 1989. "'Johnny's a Gentleman, But Jimmie's a Mug': Press Coverage During the 1930's of Myrtle McGraw's Study of Johnny and Jimmy Woods." *Journal of the History of the Behavioral Sciences* 25:356–636.

Elkind, D. 1986. "Formal Education and Early Childhood Education: An Essential Difference."*Phi Delta Kappan* 5:631–636.

Hansl, E.V.B. 1935. "Incredible Twins." *Parents Magazine* (May):55; 57.

Hetherington, E.M. and Parke, R.D. 1993. *Child Psychology*. New York: McGraw Hill, 4th ed.

Literary Digest 1933a. "Johnny's a Gentleman, But Jimmie's a Mug." 18:17.

_____. 1933b. "Conditioned Infant Excels Twin Brother." 18:18.

McGraw, M.B. 1935. *Growth: A Study of Johnny and Jimmy*. New York: Appleton-Century-Crofts.

_____. 1939. "Later Development of Children Specially Trained During Infancy: Johnny and Jimmy at School Age." *Child Development* 10:1–19.

_____. 1940. "Signals of Growth." *Child Study* 18:8–10; 31.

_____. 1942. "Johnny and Jimmy." *New York Times Magazine* (April 19):22–23; 37.

_____. 1958. "Infant Motor Development: A Study of the Effects of Special Exercise." Paper presented at Soviet Sports Conference.

_____. 1986. "The Problem of Using Secondary Sources: Elkind's Blunder." Unpublished letter submitted to *Scientific American* (July 23).

Munn, N.L. 1961. *Psychology*. 4th ed. Boston: Houghton Mifflin.

Nelkin, D. 1987. *Selling Science*. New York: W. H. Freeman and Company.

New York Times. 1934a. "Earliest Training not Retained by Infants." (September 9) 2:7

_____. 1934b. "Conditioning the Child." (September 16).

_____. 1935a. "Twins in Uniform." (January 1).

_____. 1935b. "Conditioned Child Proves Superiority." (January 16).

_____. 1937a. "Twins, 5, are Bored with Interviewer." (April 16).

_____. 1937b. "More Child Study." (November 24, Section 4).

_____. 1938. "Brother Bosses Scientific Twin." (April 16).

_____. 1939a. "Scientific Twin Hates School; One That Just Growed Loves It." (April 17).

_____. 1939b. "Scientific Twins Find Circus Dull." (April 19).

_____. 1941. "Unconditioned Twin, at 9, Makes Scientific One Seem Tongue-Tied." (April 17).

_____. 1942. "Twins React Alike to Circus Thrills." (April 19).

Razel, M. 1988. "Call for a Follow-Up Study of Experiments on Long-Term Deprivation of Human Infants." *Perceptual Motor Skills* 67:147–158.

Senn, M. 1972. "Interview with M. McGraw." Hastings-on-Hudson, New York. (May 9). *Oral Histories of Child Development: 1962-1983* Milton Senn Collection at the History of Medicine Division, National Library of Medicine, National Institutes of Health, Bethesda, MD.

Shaften, D.R. 1993. *Developmental Psychology*. 3rd ed., Pacific Grove, CA: Brooks/Cole

Sherbon, F.B. 1938. "Review of a Study on Growth." *The Medical Woman's Journal* (November):1–7.

❧ 3 ❧

Later Development of Children Specially Trained During Infancy: Johnny and Jimmy at School Age

Introduction

In 1932 Johnny and Jimmy, twins, became the subjects of an intensive study of behavior development at the Normal Child Development Study of the Babies Hospital, Columbia-Presbyterian Medical Center. Before the study had been under way for two years, it attracted wide interest because of the reports of a baby less than a year old swimming with his face under water, ascending steep inclines and by the time he was sixteen months old moving around with considerable skill on roller skates. These were the achievements which gave the study a popular interest, but such performances were incidental to the primary objectives which were (1) to analyze the sequential phases of changes through which a growing infant passes in the achievement of a given performance and (2) to determine whether these phases are altered by certain prescribed conditions, viz., the daily stimulation of activity on the one hand and the restriction of activity on the other.

Such an investigation was timely because of the current pediatric and psychiatric notion that infants should not be over-stimulated and because of a general assumption following a co-twin study by Dr. Gesell (1929) that the immature nervous system of the infant is unresponsive to practice effects and that development during infancy is essentially a matter of neural maturation. It is proverbial that among older children and adults practice leads to improvement in performance. If the infant is unresponsive to practice effects and if, on the other hand, the adult is responsive, it is a reasonable as-

Reprinted from *Child Development*, 1939, 10(1):1–19 by permission from The Society for Research in Child Development, Inc., and Mitzi Wertheim.

sumption that there must come a time in the course of the child's development when improvement by virtue of experience begins. It therefore seemed
to the writer that if one began soon after birth to stimulate one member of a
set of twins in certain activities and within certain limits to restrict the activities of the other, it might be possible to ascertain that period in life when
the individual begins to profit by experience or repetition of performance.
Johnny and Jimmy were selected as the subjects for this investigation and a
group of 57 infants examined at weekly or bi-weekly intervals served as
controls. The general procedure during the first twenty-two months old was
to observe both babies at the laboratory five days a week from nine until
five. During this time Johnny was stimulated daily to engage in activities to
the extent of his capabilities, whereas Jimmy was left with a few toys unhin-
dered in his crib except for disturbances accompanying routine care. When
they were twenty-two months old, Jimmy was given a period of two and
one-half months of intensive practice in those same activities in which
Johnny had been given earlier and more prolonged exercise. A detailed report of this investigation was made in 1935 in *Growth: A Study of Johnny
and Jimmy.*

In this report it was pointed out that there was no one age period or developmental stage which clearly demarcates an earlier state of immaturity during which the child is incapable of improving through practice from the subsequent state in which improvement through practice becomes feasible. The
impossibility of identifying such a critical period in the development of the
individual results, it would seem, from the fact that the nervous system does
not mature uniformly. There are critical periods dependent upon the maturational status of the nervous system, but these periods vary with respect to
the particular activity under consideration. Before training or practice can be
economically provided, it is essential to determine, by the observation of behavior symptoms, the periods of greatest susceptibility for each type of activity. It will be recalled that while Johnny was induced to roller-skate with
considerable skill by the time he was sixteen months old, repeated daily
practice in a seemingly simple activity like tricycling evoked no improvement
either in technique or performance until he approached nineteen months of
age. The discrepancy between these two activities affords a striking example,
because under ordinary conditions a child learns to tricycle much younger
than he does to roller-skate.

It has been four years now since this special study of Johnny and Jimmy
was discontinued. Although they have at specific intervals during these four
years been given follow-up examinations at the laboratory, their life at home
has otherwise been comparable to that of other New York children of their
socioeconomic status. However, an immediate question arises as to the se-

quelae of their contrasting experiences during the first twenty-two months of life. Now that they are just attaining school age, when a new chapter of a child's life is opened, it is desirable that we consider their relative development to date in the light of their earlier experiences.

At the time of the original report, before there had been an opportunity of observing the subsequent influence of special exercise, the writer commented,

> The permanency of the expansion which an action-pattern gains through additional exercise is contingent upon the degree of fixity the behavior-pattern had achieved at the time the modifying agent, i.e., the factor of special exercise, was withdrawn. It does not necessarily follow that a performance which has been developed under special conditions will be retained after those conditions are removed. Unless the behavior-pattern has become fixed it is only reasonable to expect that there will be a loss in performance when the conditions which brought it about are discontinued. Correspondingly, in the growth of a behavior-pattern has been hindered through restriction, it is to be expected that recovery will be evident when the restrictions are removed(McGraw 1935:310–311).

The early investigation brought out the fact that certain activities of infants can, through exercise, be brought to a much higher level of achievement than was normally assumed, but it will require new and longer studies to determine the extent to which practice should be enforced in order to render its results comparatively permanent. It is well known that many adults who learn a performance such as bicycling in childhood can, after a lapse of years without practice, pick it up in fairly short order. On the other hand, an adult resuming his attack on a language which he has not spoken since early childhood will show more than an initial awkwardness in re-learning the language although he may acquire it with greater facility than someone who had never been exposed to it before. In other activities an adult often appears to be a virtual tyro despite his childhood accomplishments in a given field. An adult's loss of childhood skills in marble shooting, ball throwing, etc., is such a common occurrence that it is a familiar subject for the cartoonist and other humorists. General conclusions concerning the permanent effects of practice are therefore impossible since different types of skills exhibit wide variation in their tendency to be retained or lost.

In evaluating the permanency of practice effects with Johnny and Jimmy, we shall consider first those laboratory activities in which Johnny had been given long and intensive training during infancy but in which Jimmy had received only two and one-half months of intensive training beginning when he was twenty-two months old. During the past four years, they have been examined in these same performances at intervals varying from two weeks to six months.

Laboratory Studies

Tricycling

One of the most interesting attainments of Johnny and Jimmy during their period of intensive study was that of manipulating a simple tricycle. The initial practice period was begun with Johnny when he was eleven months old. It was pointed out in *Growth* that this activity was initiated before his neuromuscular mechanisms were ready for such a performance, as it was eight months before he began to show distinct comprehension of the situation (McGraw 1935:255). After he showed some degree of mastery he acquired an easy, skillful performance in about two months. It was also pointed out that Johnny had apparently suffered by his long and futile practice periods. Jimmy, whose training began when he was twenty-two months old—presumably when his neuromuscular mechanisms were in a state of readiness—accomplished an easy performance in a shorter time than did Johnny. After the period of intensive training was discontinued, both boys were observed at the laboratory in their tricycling behavior, and at no time did either of them show any distinct loss in this particular skill. The twins were living at home during this time. Although they did not possess a tricycle in their own home it was impossible to ascertain whether they had access to one in the nursery school or on the playground. On the other hand, it may be that tricycling, like bicycling, automobile driving, etc., was a type of skill which does not deteriorate appreciably through lack of practice once it has been definitely acquired.

Skating

In some of the other gross motor skills we find a different story. It was pointed out in the earlier report that Johnny, who began his roller skating experience when slightly less than a year old, enjoyed an advantage over his twin, whose practice in this activity was begun at the age of twenty-two months. This advantage appeared to be attributable to several factors. In the first place, from purely mechanical considerations, a child who is just learning to walk possesses because of his relatively wide base and short legs greater static equilibrium than does the child who is older and therefore taller (McGraw and Weinbach 1936). In the second place a child who is just beginning to walk, has ample experience in falling, less distance to fall, and is therefore less disturbed when he falls on skates than is the older child. Furthermore, a child's achievement in a particular skill like skating is determined to some extent by the number of distracting or interfering interests which play upon a performance at a given time. The toddler of a year was less responsive to the world about him than is the child of two years or older. Johnny was at the threshold of independent walking when the skating practice was initiated and apparently experienced an advantage in static equilibrium. Also, having had less general experience at the age of twelve months, he was less hindered by distractions and emotional factors and Johnny therefore could exert himself to the limit of his neuromuscular abilities, whereas Jimmy's performances at the age of two years were modified considerably by interference from broader interests, so that his activities at the time were not the result of optimum neuromuscular coordinations. Even after a practice period of two and one-half months, Jimmy had still not acquired a well-integrated skating movement.

The children were given skates as a present shortly after their daily laboratory visits were terminated, but these skates were demolished within a few weeks and their parents report that they have had no other skating experience except for the follow-up laboratory tests. Their performances on these occasions corroborate the parents' report. In brief summary of the notes which have accrued on their skating behavior during these four years, it may be said that soon after their practice period both children began to show a loss in skill and disorganization of the skating pattern. When the children were about three years old, it was noted that the chief source of their difficulty was in maintaining the erect posture when on skates. The aspect of balance had undergone greater disorganization than the actual progressive movements. Disorganization of the skating behavior was more pronounced in Johnny's performances than in Jimmy's. This difference, however, appeared to be due to Johnny's attitude of abandon and a tendency to lunge forward even though he had lost equilibrium, whereas Jimmy adhered more to the short, stiff strokes characteristic of his early practice period. Despite this difference both children began to show an increased tendency to lose their balance and tumble, when compared with their earlier performances. Neither of them skates well today.

The fact that the children showed no loss of skill in tricycling but complete disorganization of roller-skating raises the question as to why certain skills deteriorate through lack of exercise and others do not. The answer to this question is, of course, problematical since the factors which control a growing behavior are multiple and were not in this study controlled by laboratory measurement. However, from observations of other children in our study as well as these particular boys it seems clear that at least three factors play an important role in determining the permanency or deterioration of a motor skill when no special exercise of the function has occurred during a period of years. The first factor is the lack of practice; the second is the influence of the child's attitude toward the performance, and the third is the changing configuration of the bodily structures as a result of physical growth. It is impossible to evaluate the effect of disuse of function without taking into account the factors or attitude and organic structures. From direct observation of the twins' skating behavior the writer is of the opinion that changes in bodily proportions were of considerable significance in the disorganization of skating behavior. Actually the child of five or six years, because of his relatively long legs and narrow base, has a set of structures with which to perform the task of skating different from those of the toddler. With the set of structures characteristic of the toddler, Johnny developed a well coordinated skating movement. Johnny's attitude toward skating continued to be cooperative and favorable during the entire four years.

There are other motor activities in which he showed no appreciable loss of skill despite the lack of exercise. It seems reasonable to infer, and certainly the character of his behavior indicated, that the disorganization was brought about not solely because of disuse of function but also because during the non-practice period important growth changes occurred in the bodily mechanisms which function in the act of skating. It is reasonable to assume that if the practice had continued he would have altered his behavior gradually to meet the new and slowly developing structural demands. Jimmy, who was less skillful in skating than Johnny at the end of their practice periods, has also shown deterioration. The fact that both began to show marked difficulty in balancing at about the same age lends support to the contention that physiological growth was an important influence in the disorganization of their skating behavior.

Slides

Another activity which seems to have undergone alteration in the course of years, though it was not so completely disorganized as roller-skating, was that of ascending steep slopes. It may be recalled that at the end of the experimental period, Johnny was able to ascend easily a slope of 70 degrees and Jimmy could with somewhat less ease, scale the incline of 61 degrees (McGraw 1935). For some months immediately following their practice period Johnny showed, on follow-up examinations, no loss in motor skill in mounting even the steepest slopes. Jimmy, on the other hand, showed some initial loss, but then began to improve both in motor skill and in persistence. Alterations in their methods of ascending the slides became most obvious when the children were about three years old. At this time it was noted that they had great difficulty in managing their longer legs. They would attempt to ascend on their knees while grasping the ridges and pulling vigorously with their upper extremities. The most outstanding individual difference was Jimmy's persistence in trying to ascend in this manner while Johnny, after a few trials, would shift to the more efficient method of his toes. He also climbed the steep slopes with greater ease and muscular coordination. He gives the impression of having better gripping power in his toes than Jimmy or most six year old children whom we have had occasion to observe in this activity.

At the present time the writer is less impressed as to the extent to which daily practice might have functioned in Johnny's achievements on the slopes than she was at the time of the original report. Practice undoubtedly operated in his ascending slopes of 61 and 70 degrees respectively. Other infants in our laboratory however, have ascended slopes as steep as 40 or 46 degrees without great difficulty even when they had not been given systematic practice in such performances. To the experienced observer, it is obvious that infants can more easily ascend slopes of this order than can older children. The advantages enjoyed by the infant appear to be due somewhat to differences in body configurations. The center of gravity in the infant's body is relatively higher and his legs are relatively shorter. It is therefore possible for him to get his chest, and thereby his center of gravity, nearer to the slide without raising his pelvic girdle too high. It is also possible that the texture of infant skin was such as to create a higher correlation coefficient between his hands and feet and the underlying surface. Whatever the reasons may be, the fact remains that the babies can stick to and ascend these slopes with greater facility than the older children. The difficulties of older children are manifested by slipping or a deficient gripping power in the toes and in managing their long legs. Alterations in the slide climbing behavior of Johnny and Jimmy during the past four years were of a similar nature. It was a reasonable inference, therefore, that growth changes in body proportions played a large role in altering their method of ascending slopes. Any permanent effects of the early practice which Johnny enjoyed are general, as indicated by superior motor coordinations, except for the comparative readiness with which he shifted from an inferior to a more effective method of managing his long legs and using his toes.

In descending these slopes neither Johnny nor Jimmy at any time showed any distinct loss of motor skill or alteration of method. During the first few months following the practice period, Jimmy was more timid and required greater urging to descend, but after the first three or four months, he overcame this hesitancy and manifested no lack of motor skill subsequently.

Getting off Pedestals

Post-practice behavior of the children in getting off pedestals is similar to that of descending the slides, since no definite deterioration of motor skill was manifested by either of the boys. The pedestals ranged in height from 14-1/2 to 63-1/4 inches. At no time during the four-year interim has Johnny shown any loss of skill or hesitancy in getting off these pedestals. At the time the practice period was terminated Jimmy was easily and deliberately climbing off the 63-1/4 inch pedestal. During the months immediately following his two and one-half month practice period, there appeared a regression in his general attitude or emotional adjustment to the situation which interfered with his motor performance. He would complain even on the lower pedestals and would refuse to get off the higher ones. Behavior of this character continued more or less until he was about four and one-half years old when his attitude seemed to change to cooperation and some enjoyment of the situation. The failure of Jimmy to get off the higher pedestals was not so much a deterioration of the motor coordinations formed during his practice period but rather a recurrence of attitudes established prior to his practice period. Once he could be induced to climb down he could do so with considerable ease. He has never, however, attained the agility which Johnny manifests in this performance. At the present time the individual differences are indicated not so much in terms of their achievements as in the degree of coordination which they show when performing the same act. This superior motor coordination on Johnny's part is reasonably attributable, in a measure, to the difference in amount of exercise the children received in the activity during their first two years of life. Climbing off pedestals is another type of performance, the motor aspects of which do not show appreciable loss through lack of exercise once the motor habits are well established.

Jumping

In the report of early behavior of the children in the jumping activity (McGraw 1935) it will be noted that in this type of performance the attitude of the child played a major role in determining the somatic response. At the end of the experimental periods the investigator felt not only that Johnny had established a cooperative attitude, but that the integration of the essential movements in jumping had been accelerated. Jimmy, on the other hand, even after two and one-half months of daily exercise could not be induced to jump off a tall pedestal into the outstretched arms of an adult. It seemed unquestionable that Johnny experienced an advantage in this performance by having had practice in the activity before his increased perceptive and emotional capacities added complexities to the somatic or motor aspect of the behavior. During Jimmy's two and one-half months of practice (at two years of age) his attitude became definitely more acquiescent, but the alteration of attitude was not sufficient to effect an integrated jumping performance. At the end of the exercise periods, when they were 26 months old, Johnny was gleefully leaping from tall pedestals with even a slight "spring" as he threw himself forward. Jimmy, happy enough, would stand on the pedestal, shifting his weight from one foot to the other, squatting, and in other ways indicating his urge to go forward, but was not quite able to consummate the performance. During the first few months after their practice periods were terminated, Johnny began to show a less cooperative attitude and for a while there was even less motor skill in his action, that, when he did jump there was noticeably less grace in his movements. This loss or deterioration was, however, temporary and intermittent, appearing at intervals over a period of six or eight months. After that relatively brief period, jumping was again one of his most enjoyable activities, and he subsequently showed no distinct loss in motor coordinations. It appeared to the experimenter that this early change in attitude was due in part to imitation of his brother's behavior rather than to real hesitancy in carrying out the act.

For some months after their practice periods terminated Jimmy showed a definite regression in his behavior. He would cling tenaciously to the adult, refuse to stand up on the pedestal, and cry lustily. When he was about three years old, however, his attitude became more acquiescent, and his behavior comparable to that manifested during his exercise period. He would stoop or squat on the pedestal, reach toward the adult, and if the adult was only about eight or ten inches away he would throw his shoulders forward. He has steadily shown improvement and is now willing to jump, but he has not manifested the abandonment nor been willing to jump as far as Johnny. In this activity there has been a definite residual of the early training period in Johnny's favor.

Purposive Manipulation of Graded Stools

Growth offers an analysis of the children's development in manipulating stools of various heights in order to obtain lures which had been placed out of reach (McGraw 1935:174–183). At the time the experimental practice period terminated, neither of the children had attained the maximum degree of proficiency in this activity. At that time Johnny would juggle eight different pedestals, ranging from 7 1/2 to 63 1/4 inches in height, in order to climb up and obtain an object hung some nine feet above the floor. However, he still had a tendency to push all the stools in a cluster, not seriatim, usually with the tallest one beneath the lure. On a few occasions he had shown some tendency to eliminate unnecessary pedestals, but this aspect had not become a fixed part of his behavior activity when the practice period ended. Jimmy, on the other hand, was, at the same time, able to manipulate two or three stools successfully in order to obtain the lure placed upon the 63 1/4 inch pedestal, but he had at no time successfully obtained the object hung from the ceiling, which arrangement demanded his making use of the taller pedestals for climbing. Johnny's practice period began in this activity when he was eighteen months old, and Jimmy's when he was twenty-two and one-half months old. It was pointed out in the original report that the time span between the inception of the practice periods for the two children was not great and therefore came nearer to striking the critical period for both children, i.e., that period when development in this type of activity would be most susceptible to advancement.

It would require too much detail in order to delineate the various changes in their performances in this situation during the four years they have been returning to the laboratory for follow-up examinations. It is sufficient to state that at the present time, as a rule, Johnny and Jimmy both arrange the pedestals purposively to obtain the lure when it is hung at least nine feet above the floor. Both children carefully discriminate in placing the tallest pedestal under the object. Johnny, however, carelessly and dexterously pushes all the other seven pedestals in a cluster about the tallest one with no definite order of arrangement. He then goes clambering up two or three of the stools, as may be necessary, for him to gain the top of the tallest one. He shows no hesitancy or difficulty in bridging wide gaps in order to pass from one stool to another when they are not in juxtaposition, and he shows no hesitancy in standing erect on the tallest stool in order to reach the lure. It never seems to trouble him that he has pushed all the pedestals in a cluster although he makes use of only two or three for the purpose of climbing.

Jimmy, in contrast, works diligently arranging the stools in a graded sequence with the tallest one carefully placed under the object. Once he has them arranged in a stairway he will climb up and stand on the one next to the tallest pedestal, but it is only rarely that he can be induced to climb up on the tallest one so as to obtain the lure. He has definitely less courage and less motor coordination in handling his body. It is the writer's impression that Johnny's failure to arrange the pedestals in order according to their relative height was a residual of his habits established earlier when his discrimination was inadequate to allow him to do so. The fact that it is not lack of discrimination at the present time was brought out one day when he was shown some movies of his performances taken before he was two years old. This particular reel included an occasion when he had arranged most of the pedestals in stair formation. Johnny's remark was, "That isn't me. I didn't put them straight like that." With his greater agility he doesn't feel the need of an orderly arrangement, whereas Jimmy, who was more cautious in climbing, tries to make that aspect of the situation as easy as possible. The impression that the tendency of Johnny and Jimmy to employ all eight stools, even though three properly selected ones would have sufficed is a residual of their early practice in this situation was substantiated by the behavior of a small group of children from six to eight years of age, who had never been exposed to this particular situation before. At this stage of maturity these children would make use of only a few of the pedestals, even if they were not properly chosen. In other words, they had developed beyond the stage of thinking an additional one would help solve the problem.

Purposive Manipulation of Graded Boxes

This situation called for the stacking of two or more boxes of different size on top of each other in order to obtain a lure suspended from the ceiling (McGraw 1935:153–190). When the practice periods were discontinued, Johnny was skillful in stacking three boxes in order and had occasionally successfully stacked four boxes, but he was actually lacking in physical height to stack the fourth box easily. Jimmy, by contrast, at the same chronological age and after a period of two and one-half months practice, had not arrived at the stage of placing one box on top of another for the purpose of climbing up to reach the lure. Therefore in this situation neither child had attained his maximum development in performances of this order, though Johnny was considerably advanced beyond Jimmy.

In the four year interim during which they have been given follow-up examinations in this situation Johnny has shown many fluctuations, the details of which would be too laborious for the reader if reported here. Jimmy was three years old before he began placing one box on top of another, and it may be that this is about the age when children would normally begin to engage in activities of this order if they were not given specific stimulation to do so. At the present time both children usually pile all four boxes on top of one another in order to obtain a suspended lure. Jimmy is more careful in arranging the boxes in order according to size, that is, with the largest one on the bottom and the smallest one on the top. He often shows a reluctance to climb up after he has neatly arranged the boxes. Johnny, who is careless and less discriminating in the order of arrangement, shows greater skill in climbing and maintaining his balance on even a quite unstable structure. It would seem that in this particular situation Johnny's advanced motor skill and courage operate to make him either careless or less discriminating of the relative sizes and placement of the boxes. Again, Jimmy, who has less courage in motor performances, is more meticulous in making arrangements, so that the motor aspect will be as easily accomplished as possible.

When a long range view is taken covering the twins' behavior in these several laboratory situations, during the past four years, it is clear that there are at least three different situations in which Johnny showed no loss of proficiency and in which any loss which Jimmy might have shown could be attributed more to his emotional or attitudinal status than to motor inadequacy. These activities are (a) tricycling, (b) getting off pedestals, and (c) descending slides. When we examine these activities we are struck with the fact that their mode of performance has suffered no major alteration. In other words, so far as the motor aspect of these performances goes, the twins had achieved a high degree of integration or maturity at the time the practice periods were terminated. Given relatively the same heights and slopes and a tricycle of relatively the same size, the two year old and the six year old child tend to get off pedestals or go down slides in essentially the same manner, and also to propel a tricycle in the same fashion.

In at least two different activities, and in certain aspects of a third, the mode of performance seems to have changed primarily because of alterations in the boys themselves as an aspect of physical growth, specifically because of their relatively longer legs and the shift in their centers of gravity. These alterations are especially noted in skating and in ascending steep slopes. The influence of changes in bodily growth upon behavior was also observed in the way in which the boys manipulated graded pedestals to obtain suspended lures. Actually, because of his short stature, the two year old child who strikes the tall pedestals nearer their base is less liable to tip them over and can, therefore, move them with greater ease than can the six year old, whose contact point in pushing strikes the pedestal nearer its center of gravity.

Purposive arrangement of pedestals and manipulation of boxes were activities which were probably in the most fluid state at the time the special practice periods were terminated. For that reason the performances of both children have shown greater fluctuations in these situations than in activities which were more stabilized. While the achievement difference of the two boys was great when special practice was abandoned, their relative efficiencies have, during the past three years, approached each other though the two children adopt somewhat different methods in demonstrating their merits. As might be expected there is comparatively less diversity in their present achievements in those activities which were not so stabilized when practice was abandoned.

Jumping from a tall pedestal seems to stand in a class by itself. It appeared that through practice the motor aspects of this performance had been greatly accelerated in Johnny. However, his performance had reached a high degree of fixity and suffered only a temporary loss immediately following the practice period, whereas, the attitude which prevailed with Jimmy during his period of isolation had become so fixed that the two and one-half months of practice he experienced when he was about two years old were inadequate to counteract it. With increasing maturity Jimmy's emotional disturbance was somewhat abated, but it has never become sufficient to eliminate obvious muscular tension in his performance even though he will complete the jump.

Swimming

No follow-up study has been made of the swimming behavior. It will be recalled that the original investigation of this activity (McGraw 1935:122–136) terminated when the children were seventeen months old. During the ensuing four years there have been three or four occasions when their aquatic behavior was observed. Until the age of six years Johnny showed noticeably greater fortitude in jumping into the water and playing about, though he was not able and was not urged to engage in independent swimming. However, on the last occasion they were taken to a lake, when they were a little more than six years old, Jimmy was quite brave about jumping into the water with a ring around him for support, whereas Johnny showed some inhibition about doing so. Just why this change in attitude should have taken place is a matter for conjecture. An interesting difference was noted in their swimming strokes. When supported by a ring, both would venture out beyond their depth, but Jimmy maintained a vertical position as he made rapid treading movements with his legs. Johnny, on the other hand, assumed the horizontal position and engaged in graceful crawl strokes of the more advanced swimmer. Just why he should have taken up a method of movement which by experts is considered to be more efficient than the natural dog paddle of beginners, is beyond present explanation as no attempt was ever made to teach Johnny strokes during his period of infantile swimming and at that time he used the typical dog paddle. It is unfortunate that this behavior could not have been studied further, especially since it appears that this is another activity in which the body proportions constitute a large influence in determining the manner of behavior. An infant, whose head is relatively heavier with respect to his total body, cannot maintain his face above the water level. Beginning swimmers between the ages of four and six years have a tendency to maintain a vertical position, if allowed artificial support, and their first movements without support are of the struggling order.

INTERPRETATION

In the light of their subsequent behavior, both in and out of the laboratory, it is safe to state that at the present time Johnny usually manifests greater motor coordination and daring in physical performances. Jimmy, who is more awkward and timid, exercises devious methods of rendering more easy the motor aspect of a given activity, so that both children may end up with the same result in terms of final accomplishment.

It has been pointed out that the degree to which an accelerated activity may retain its advanced status after the modifying factor of intensive exercise has been withdrawn is proportional to the degree of fixity the behavior pattern had attained when practice was discontinued. Fixity in this connection means whether the manner of performance had stabilized into a well integrated movement. Examples of well stabilized activities which manifested no major alteration through *lack* of practice are tricycling, getting off pedestals and descending slopes. Manipulating pedestals and boxes was in a fluid state, and their subsequent performances have shown many fluctuations.

Fixity in mode of behavior, however, is not the only factor determining the permanency of acceleration. Another important factor—viz: organic change within the individual—has been recognized only by the analysis of later performances in skating and ascending slides. Johnny's mode of performance in these activities when he was two years old was highly skillful and of an order comparable to that of the efficient adult engaging in a similar activity. If one considered the activity in the abstract, one would say that his skating and slope-climbing behavior had attained a high degree of fixity. Yet these activities suffered alteration, or loss in skill. This loss of skill appears to have been due primarily to change in bodily proportions. Thus it seems that a child may, in a given activity, attain a high degree of skill and his manner of operation may be the most approved or efficient with the organic structure he has to work with at the time. If, however, at a later age he has a different set of organic structures for performing the same activity, then his behavior may show disorganization characteristic of the novice. Retention of general muscular coordination may persist but the actual patterning of the particular skill or behavior will be interrupted until the child has learned to operate the new set of body mechanisms. The comparative behavior of Johnny and Jimmy in roller-skating illustrates this point. Even though the activity became disorganized in both children as they grew taller, superior general motor coordinations could be detected in Johnny's movements.

Johnny's superior motor coordination is evinced not only in adaptive skills as mentioned above but also in the common, more organic motor movements, such as the assumption of an erect posture, walking and falling. Quantitative determinations of their efficiency in walking were ascertained periodically at six month intervals from the time they were about two years old. While these measurements are too scattered to justify elaborate treatment, they do show that Johnny was consistently a little ahead of Jimmy, and to that extent these objective measurements corroborate the observational data.

In summary it might be said that the alteration or deterioration of performance, particularly motor performances through lack of exercise or practice over a long period of time, is determined (1) by the state of maturity the activity had attained at the time the practice was withdrawn, (2) by alteration in body mechanics or other physical changes which necessitate an alteration in the form of the behavior pattern, and (3) by change in emotional or attitudinal adjustment which operates as an inhibiting or facilitating factor in a particular performance.

Attitudes

In evaluating the influence of attitudes upon somatic performances we have ordinarily considered only the attitude prevailing at a moment and concomitant with a particular motor behavior. It is well to consider also the persistence of a characteristic attitude which seems to predominate from day to day when the child is confronted with the same situation. That is, if a child week after week is placed on a pedestal and requested to jump to an adult and if on each occasion he stoops, clings to the adult and cries, then it may be said his characteristic attitude for the period of time is one which inhibits the motor performance of jumping. If however, after a period of months he begins to show more fortitude not only in the jumping situation but in all comparable activities, then the possibility of general increase in emotional control warrants consideration. More specifically the older child, because of his physiological maturity, is capable of attitudes and adjustment which could not be evoked from the infant or younger child. It was pointed out in

Growth that an attitude, at a given moment, may arise from some developmental imbalance of the neuromuscular system (McGraw 1935:287). For example, the baby who pulls himself up beside the bed and stands crying lustily may do so because he has not developed the concomitant coordination of letting himself down to the sitting position. The attitude arises from the organic status of his neuromuscular development and not from some personal disagreeable experience associated with that situation. Likewise, it would seem, a change in attitude may be brought about not through direct and individual experience in a specific situation but through the integration of other aspects of development i.e., as a result of general physiological development. It is common observation that the period of greatest irritability, negativism, temper tantrums, etc., is around twenty-four to thirty months. These common observations are substantiated somewhat by the investigation of Reynolds (1928) who shows that there is a diminishing manifestation of negativism after two years of age. Anderson (1932) interpreting Olson's (1930:28–30) study of "Problem Tendencies in Children," regards the period between two and four years as one in which "a fanning out of emotional life" takes place. It seems evident from Anderson's discussion that he has in mind not only diminishing emotional flare-ups in the older child but increasing differentiation and specificity of emotional expression. The older child gets angry at things which should make him angry, while it is common observation that the younger child may go into a tantrum when apparently everything is being done to appease him. It is also common knowledge that the intellectual and motor abilities of the two year old child have not attained equilibrium.

It seems reasonable to ascribe this characteristic and trigger irritability of the toddler to the many developmental behavior potentials which he has acquired during the first two years of life, but which have not been organized into adequate social responses. When these behavior potentials become integrated then the child achieves a degree of emotional maturity regardless of his personal experiences in particular situations. In other words, emotional maturity may arise from increasing integration of intellectual and motor development as well as from direct personal and conditioning experiences. These statements sound tautological. It is difficult to make them otherwise since the different aspects of a behavior activity are interdependent and mutually influential. Illustrations may add some clarity to the analysis: A child who is just beginning to walk may leap off a tall pedestal without hesitancy because he has not developed adequate powers of height discrimination to appreciate all the possibilities of the situation. A little later, when his powers of discrimination have advanced, he squats and refuses to jump. Again it must be emphasized that his behavior has changed not because of an unfortunate personal experience but because of development in another aspect of his behavior. At this time he has adequate motor mechanisms for jumping but he has achieved powers of discrimination and for a time the one functions to inhibit the other. Subsequently when the two types of development become organized and well integrated he will again leap from the pedestal, and his attitude is acquiescent or cooperative not for lack of discrimination, but because he can exercise his motor functions in terms of his discrimination.

In a discussion of the gross chronological periods when the most rapid development occurs during the first two years, the following categories were suggested in *Growth*:

> During the first four or five months the behavior of the baby is still at an infracortical level. Therefore, the greatest increment of development in behavior at this time consists of a gradual recession of some overt primitive reflexes as behavior-patterns of a higher order begin to emerge.
>
> Growth during the second period is shown largely by an ever increasing control of the individual infant over motor activities in the upper part of the body, that is, in the region of the head and shoulder girdle.
>
> During the third period the greatest increments in behavioral development are observed in the motor activities involving the lower part of the body—in the region of the pelvic girdle and lower extremities.
>
> During the fourth period great strides are evident in the development of an understanding of relationships—associational, conditional, and symbolic of retention and recall of these relations, of comprehension and use of language, imitative tendencies, etc. (McGraw 1935:192–193).

If a multitude of behavior potentials have been acquired during the first two years, it is a reasonable inference that in the immediately following years rapid development will be in the nature of integration or organization of these more or less independent behavior potentials. A concomitant development of this integrative process will be an increased degree of emotional stabilization. General observations of children's behavior support this hypothesis.

A study of the notes on Jimmy's attitudinal adjustment to the laboratory situations during the last four years indicates that a general maturing began to take place when he was about three years old. During the first year after they left the laboratory his attitude on return visits was infantile, comparable to that manifested during his period of restriction. He would cling to the experimenter, cry and refuse to go through with many of the performances for which he obviously had adequate neuromuscular development. At about three years of age, he began to show a more interested and cooperative attitude; the change was shown in practically all the laboratory situations. His motor performances began to improve and have continued to improve to the present time.

Since no definite effort was made during this time to re-educate his attitudinal adjustments, it is assumed that the change was a result of gradual development in understanding and integration of his behavior in general. Jimmy's responses both in and outside the laboratory are more representative of the typical child of his age than are Johnny's. This period of development in emotional and social adjustment achieved by Jimmy parallels the period for such development in children in general as ordinarily observed.

One attitudinal difference in the two children which was marked during their experimental period persists even today, though the difference is less marked than it was when they were about two years of age. Jimmy has always been more concerned with end result and Johnny with *modus operandi*. It cannot be said that Johnny's greater interest in the way things work rather than in obtaining lures arose from the more extensive laboratory experience. It is however reasonable, from the very nature of the laboratory experience, to assume that it was not without influence in developing such an attitude. Here he learned to work for the sake of performance even when the lure had no particular appeal. The fact that this difference in their interests was more marked during the experimental period than it is today would in part, justify the opinion that the laboratory experience influenced Johnny's development in this respect. At the present time a sort of rivalry exists between them, but it is counter-balanced by their spirit of comradeship. In the main, they appear to be less sensitive to failure than most children of their chronological age who have come under the observation of the writer.

Personality Development

In view of the importance attached to the early experiences of life in determining later personality makeup, and in consideration of the rather popular notion that over-stimulation during infancy forecasts a neurotic child, two attempts were made to evaluate the personality and social adjustments of the children as of six years of age. Dr. William S. Langford, through psychiatric interviews with the mother and children, arrived at the following evaluation of the children's emotional and social adjustments as of today:

> It is difficult to discuss these two children without stressing the differences between them and without speculating as to possible causes of these differences. Johnny and Jimmy both are regarded by the mother as happy children who benefited by their experiences in the Normal Child Development Clinic. There are no particular difficulties in home management with either of the boys. The mother believes that both, as a result of the clinic experience, are able to meet with people and adjust better in social contacts outside of the home than do her other children. When playing with the older siblings, they both are able to adapt their play to the situation; when playing with the girls it is dolls, and when playing with the boys it is guns and cowboys; they are equally happy at each type of play.
>
> There has been a difference in the attitude toward the two children at home, in part coming out of the fact that Johnny was the 'subject' in the clinic. This dif-

ference in attitude would seem important in the genesis of some of the differences between the two boys. Although the family tried to realize that the selection of Johnny for the conditioning experience in the clinic was not because of
any favoritism, it was difficult for them not to favor Jimmy and not to try to
make up for some of the things he did not get. During the latter part of the second year, after newspaper reports of the experimental studies had been published, they would, among other things, take Jimmy and encourage him to jump
off the icebox. However, the experiences in the home, which would seem to
have given Jimmy the greater security there, were more a result of circumstance.
In their infancy when, at the end of the clinic day, they were returned home,
Johnny would be tired and go to sleep; Jimmy on the other hand would be wide
awake and ready for play and socialization with his parents and siblings. For a
time Jimmy would strike Johnny and take away his toys; Johnny would not
seem to dare to hit back. One wonders strongly if this, too, was not a result of
attitudes in the home.

Jimmy seems to be more at ease in the home situation where he is the leader
of the two boys. He usually bosses Johnny about. Jimmy is quite apt to come
home with tales about Johnny and tell what he had done outside. The parents,
one feels, do not particularly encourage him in this activity. Jimmy wakes up
quickly and is on the go all day as a rule chattering a blue streak about whatever
comes into his mind. He seems closer to his mother; talks more with her and
likes to sleep with her. When he is put to bed with another sibling, he will frequently come into his mother's bed. He definitely prefers his mother to his father and feels that she likes him best. He is helpful about the house and likes to
assist his mother in tidying up and washing dishes. Jimmy cries easily when
things do not go his own way or if he is scolded. He indulges in mild temper tantrums consisting of stamping his feet when he cannot get what he wants. Earlier,
at about two years of age, he went through a period of severe temper outbursts
with breath-holding.

Johnny gives evidence of some tension; he has always been a nail biter, and
the mother states that he is a 'wiggler,' cannot sit still in a chair, and from time
to time displays quite restless sleep. In addition, Johnny has always been en-
uretic nightly. This is difficult to interpret as enuresis seems to be a family failing. There was difficulty in establishment of the day habit with all of the older
siblings and one brother did not stop his bed-wetting until nearly eleven years of
age. Jimmy still wets his bed occasionally although the mother did not tell of
this until some time after the initial interview. In addition, an adult member of
the family will occasionally wet his bed. The mother feels that this tendency
comes from the father's family where all the members have a tendency toward
urinary frequency and urgency. Johnny, among all of the Woods children, is a
thumb sucker; this began in early infancy and continues at bedtime even today
although it used to be more marked and occur in the daytime also. He sucks the
left thumb, and as an accessory movement pulls his own or bed partner's hair
with the other hand; usually the latter. The sucking is rather vigorous and the
accessory movement is so pronounced that most of the siblings prefer not to
sleep with him. Johnny is somewhat ashamed of this habit and was disinclined
to discuss it at first. Johnny is the quieter of the two children and rarely holds

conversations with the other members of the family. The mother believes he is 'deeper' and 'when you least expect it Johnny will say something.' He is dependent on his mother in being washed and dressed and tends to play in his bath. One feels that this slowness is more a result of preoccupation than of a desire to have the mother do these things for him as the mother becomes irritated at his slowness and finally does it herself rather than wait for him to complete it. Johnny is not thoughtful or helpful about the house, but tends to be 'destructive and throw things around.' The mother feels that Johnny has no fears and thinks that he would be better off if he were a little more cautious, especially in his attitude towards dangerous occupations, such as crossing streets. In the home, Johnny has a philosophical attitude and takes things as they come; he is not upset when he cannot get his own way. There are no temper outbursts and he rarely cries. During the past year he has shown a tendency to play with fire, but otherwise has shown no overt behavior difficulties. Both children when observed in the interview were friendly and cooperative and talked freely. Both boys, in common with two older siblings, speak indistinctly with a lisping difficulty in articulation suggestive of 'baby-talk.'

Jimmy showed a good deal of spontaneous chatter, but was quite apt to grow almost incoherent in the rapidity with which he spouted out detail after detail and leaped from topic to topic. He showed no marked preoccupations, but reacted to the questions immediately without taking thought as to how he answered. He talked a good deal of Johnny and showed a warm affection towards him. At the same time, he tended to bring out Johnny's bad points, telling of his enuresis and hair pulling and stating that Johnny is 'the bad one at home; bad at home and bad in school.' He said that Johnny wanted to be a drunken man when he grew up. He likes to play with Johnny best, and brags about how funny he is. Many of Jimmy's statements seemed to be made for effect and with a desire to produce a laugh as he would immediately contradict himself. His general attitudes and behavior seemed quite typical of an outgoing, exuberant, and quite usual six year old boy. He reacted without thinking and was responsive to environmental changes. In one interview when the examiner was weary, Jimmy reflected the subdued and more quiet atmosphere. Jimmy in his drawing drew a watch which he then cut out and pinned to himself, strutting around with evident pleasure at his self adornment. Johnny in the interview presented a quite different picture from that of his brother. He was friendly and cheerful but not so spontaneous. Attention was difficult to hold; he would be distracted by extraneous noises, but more often by passing thoughts of his own which would result in a seemingly irrelevant answer. He was thoughtful in answering questions and seemed to weigh his answers. His interviews came after Jimmy's and he seemed to feel that he should have everything that his brother had; making sure that he sat in the same chair, had his words written down, drew pictures, and took home a pencil as a gift. His attitude towards the members of the family was somewhat different from Jimmy's. He definitely prefers his mother and sister whereas Jimmy prefers the 'toughest' brother to the sister whom he soundly denounces as being dumb. Johnny is quite fond of his twin brother, but does belittle Jimmy's ideas of wanting to grow up twice; he himself would prefer to keep on growing until he became a giant. He did tell of a dream which he was careful to point out was 'make believe' in which an old witch hits Jimmy. Johnny talks of most of his difficulties quite freely, but is hesitant about mentioning the hair pulling. He shows a definite tendency towards self evaluation and self criticism which is not present in his brother's output. He also shows good imaginative ability and reveals in his stories evidence of a rich fantasy life although he will not express these when questioned directly. His drawings are rather well done and are not, as were Jimmy's, copies of something he sees, but rather a product of his own imagination.

Attitudes of both children toward the clinic were those of its being a pleasurable experience with the exception of having to take their clothes off. Jimmy, however, tended to protest that he liked Dr. McGraw better than Johnny, although both felt sure she had no preferences. In the clinic situation it would seem that Johnny has the greater security and tends to be the leader and to boss Jimmy about. His reactions at the birth of Dr. McGraw's child would tend to bear out his need to be wanted there.

In conclusion one might say that the boys present quite different pictures. Johnny exhibits in the home certain evidences or tensions; nail biting, motor restlessness, persistent thumb sucking as well as enuresis. The last symptom, however, is difficult to evaluate because of its prevalence in the family, and it is also present to a lesser degree in Jimmy. These symptoms could well come out of Johnny's lessened security in the home situation where his brother has been preferred and given more attention, and earlier, a greater amount of affectional demonstration. It would not seem that the conditioning in the clinic had much to do with these. The restlessness, nail biting, and sucking do not occur in the clinic setting, and the boy has great security in his relationship with the clinic personnel, especially Dr. McGraw. I should not feel that undue pressure for success in various accomplishments was a factor in this as the boy enjoyed his work. This latter, however, would seem of importance in giving Johnny his attitude of evaluating and looking over a situation.

Jimmy is the outgoing, helter skelter type of child who lives for the moment. Johnny is the more serious, thoughtful and contemplative youngster who looks to the consequences before he acts. Jimmy reacts mostly to external stimuli, Johnny, as a result of his more active inner life, reacts in a less direct manner to external stimuli and frequently gives the impression of preoccupation. Johnny gives the impression of being capable of weathering more serious environmental difficulties without blowing up because of his greater capacity for working things through; but once upset, he would respond more slowly to efforts for readjustment. Neither child seems to have suffered from the experimental study. Both make adequate social adjustments although in different fashions. The differences in their personalities may well be largely constitutionally determined and not entirely the result of their diverse earlier experience. However, one feels that these experiences are of importance. I should feel that the home attitudes were of great importance as well as Johnny's 'conditioning' and they do seem a little easier to evaluate.

Dr. Langford is in a particularly favorable position to make these interpretations since he knew the children during their experimental period and has had occasion to see them from time to time around the laboratory in addition to the specified psychiatric interviews. It is especially interesting that the opinion concerning the personality make-up of the two children as expressed by Dr. Langford is corroborated by an interpretation of their reactions to the Rorschach test. These tests were administered by Jane Sills and a blind analysis was made by Dr. Z. Piotrowski. Excerpts from Dr. Piotrowski's reports referring to Jimmy's record state:

> this boy probably makes the impression of a rather typical average child of his age, emotionally immature. Toward the environment he appears to react in a rather self-centered and labile emotional manner, and he seems to have a rather poorly developed inner life. His reasoning power is not above the average, and he does not seem especially observant. He appears to lack the capacity to preoccupy himself with imaginative and instructive games. Compared to his brother's record, he is childish, as one would expect a child of his years to be. His brother appears more independent in his actions, more individual, while this boy seems to be more dependent, more appealing to adults who like to play the role of protectors to young children.

Concerning Johnny's record Dr. Piotrowski writes,

> The complete lack of color responses in the presence of a good human response suggests that the child's reactions to the environment are determined by prompting from within himself rather than by changes in the outward situation. Probably, the child tends to be rather impersonal in his relations with people, and one might find a certain lack of emotional warmth in his attitude toward people. Although his reactions seem to be rapid, he is fairly well aware of his psychic experiences. Occasionally he is given to feelings of insecurity and uneasiness, and it is my impression that at such times the boy tends to be brave against his liking. Intellectually he seems to be above the average since only the intelligent among the small children tend to have a human movement response. Since the boy seems to lack the capacity for an immediate and effortless emotional adjustment to the environment, it is the intellect which carries the burden of adjusting. It is probably best to make contact with him on a rather intellectual and impersonal basis. His thinking would appear to have a common sense quality with occasional excursions into imaginative fantasies. His mental independence does not, however, seem to be characterized by negativism or aggression. On the whole, this child is rather self-confident and probably would impress one as capable of taking his future into his own hands. He probably gets along well with only a few children while his brother fits in better with any children's group.

On the basis of the child's reactions Dr. Piotrowski correctly surmised Johnny's record to represent that of the "trained twin since from the nature of Johnny's earlier experience he would have had a better chance to develop his inner abilities and the habit of self-observation for the purpose of avoiding future mistakes in well standardized situations."

In the main, these two interpretations of the personality differences of the children are in agreement. They are also in accord with the general opinion of the writer, whose observations of the children during their entire life has been both intensive and extensive. According to the writer's observations, outstanding personality differences are Johnny's stoical and rather philo-sophical attitude in disagreeable situations, and his escape by indulging in his own fantasies. Jimmy is more loquacious, is more conscious of, and bids for the approval of his audience. Consequently, he is more appealing to the casual observer. Since infancy he has looked and played the part of a clown. He is rather happy-go-lucky, and reflects his home environment more di-rectly. Johnny employs a more subtle method of revealing his feelings. It ap-pears that the less frequent contacts with the laboratory after the age of two years have been a handicap to Johnny. Apparently having heard that youn-ger twins came here, he spontaneously inquired, "Dr. McGraw, why did you need another baby when you sent me home?" When they left here, Jimmy went to a home admittedly more favorable to him; he also had less of an ad-justment to make. In some ways Johnny's experience here was poor preparation to meet the rough and tumble of a large family life. He had learned in many ways "to take it" and the reports indicate that he has had more than his share of taking. At home and at school Jimmy is rather the bully, often casting aspersions on Johnny and threatening his confidence. For all that, the affection between the two boys is warm. Despite the amount of public notice they have received, they are remarkably free from self-con-sciousness.

To evaluate the extent to which these personality differences are constitu-tional or determined by their early experimental experiences or their later home environment, is beyond achievement at the present time and will prob-ably remain beyond the scope of actual determination for all time. However, their experiences and behavior have been studied and chronicled as to detail in a way which is not obtainable for the average school child. If these reports and opinions serve a useful purpose in guiding the educational careers of these two children, they may at the same time be of even greater significance in showing the extent to which knowledge of early childhood experiences is of educational and psychiatric value in the adjustments of the adolescent or adult.

Intellectual and Physical Development

Both Johnny and Jimmy have consistently rated within the normal range on standardized intelligence tests. The Rorschach test indicates that Johnny is somewhat accelerated in intelligence. This paper is not the occasion to dis-cuss the correlation of the Rorschach with the Stanford-Binet, the Minne-sota, or other well-known intelligence scales. It may be that the Rorschach taps a quality of intellectual processes which is not measured by the items included in these various tests. There is one fairly common opinion among the laity at least which might at this occasion be corrected. Some have labored under the belief that special training during the first few years of a child's life will in some way raise his general intellectual endowment. It has been previously pointed out that the effect of training or exercise as applied in these studies was highly specific to those activities in which the child received daily practice. Certainly, whatever are the factors measured by standardized intelligence tests, they have not been appreciably altered by the different experimental experiences to which the twins were subjected.

There is one aspect of mental development which deserves mention, although it has no direct bearing upon the training program. In parlor discourse the earliest individual memory is a common topic. Psychologists are usually skeptical of reports of direct memories of experiences which have occurred during infancy. The usual explanation is that the person had heard the experience referred to and therefore could not distinguish between direct and verbal memories of it. Because of the peculiar circumstances of the twins first two years, it is possible to know precisely whether reference had been made to certain incidents in their early experiences. There are several instances in Johnny's behavior indicating remote memory. When Johnny was sixteen and seventeen months old he would skate through the tunnel joining Babies Hospital and Neurological Institute as he journeyed to the swimming pool in Bard Hall. These journeys through the tunnel were terminated when he was 17 months old, and he never entered the tunnel again until he was 43 months old, i.e., 26 months later. On this occasion he was walking with three other children. As he entered the tunnel his eyes widened and he suddenly remarked with a sweeping gesture, "This is where we go skating." This recall is interesting beyond the mere indication of memory for early experiences. At the time the tunnel journeys were ended, Johnny had no verbal means of expressing the act of skating. He recognized skates by name, but he had no word in his own vocabulary to indicate them. Here his remote memory is entirely integrated with verbal expression which was acquired at a later date. A year later, when he was four and one-half years old, Johnny was again escorted to the tunnel. On this occasion he asked "Where's the bathroom?" When told he would have to wait until he returned to the laboratory he replied, "Oh there's one way, way down there," pointing in the direction of the Neurological Institute. It was then recalled that when he was 17 months old he was being trained in toilet habits, and as soon as he arrived at the Neurological Institute during those days he was rushed to the toilet.

Another instance illustrates his incorporating into his vocabulary recollection of an experience before he had language facilities for expressing such experiences. Johnny and Jimmy were just two years old when they entered the elevator of an apartment house alone and Johnny pulled the lever, sending the elevator down to the basement. The occasion was subsequently never referred to, and other persons who have been associated with the children did not even know about it. More than three and one half years after, as Johnny sat somewhat meditatively watching the dial indicate the floors of the elevator in Babies Hospital, he quietly remarked, "Dr. McGraw, remember when Jimmy and me went in your house in the elevator and it went down boom!" These incidents are cited because they are comparable to citations of many parents in reporting the early memory of their children; and because of the peculiarity of experiences to which the twins were exposed during their first two years, it is possible to know definitely whether or not subsequent reference had been made to the situations. If the direct memory of experiences during the first two years of life are as indelible as these incidents indicate, then the weight attributed to influences imposed upon the infant and young child gains significance.

There is nothing of experimental significance in the physical history of the two children. Although Johnny was smaller during the first seven months of his life, he began at about eight months to maintain a superior weight gain as compared to that of Jimmy. This advantage in body weight he has sustained during the four years they have been at home. While Johnny began to show this superiority in weight gain the same time as he showed improvement in motor performances, there is no claim of a direct relationship between the two. Their health records compare favorably with those of other children on their socio-economic level, and their value to this study is only insofar as they indicate no serious sequelae of their experimental experiences during their first two years of life.

Summary

Studies of the performances of Johnny and Jimmy in particular laboratory situations indicate that the amount of retention of a motor performance, once the factor of repetition has been reduced or abandoned, is contingent upon the state of fixity the activity had attained at the time the practice-factor was withdrawn. Activities which have attained a high degree of integration may be appreciably altered if the body mechanisms are so modified through growth as to introduce new structures or elements into the situation. The natural and gradual maturing of emotional or attitudinal factors seems to influence appreciably the somatic or motor performance in particular activities.

In general endowment the two children have consistently fallen within the normal range as measured by the standardized intelligence tests. There is no reason to believe that exercise in special activities will accelerate mental functions as measured on standardized scales. There seems to be a superiority of general muscular coordination on the part of Johnny, who received the longer and more intensive practice in motor activities. In general personality make-up, Johnny also appears to be more complex as indicated both by psychiatric interviews and by analysis of responses to the Rorschach ink blots. There is no way to evaluate the extent to which their early experimental experiences operated in determining their respective personality components.

The two boys today present the picture of lively, normal six year old children who show no deleterious sequelae of the different regimes to which they were subjected during the first two years of their lives.

⮜ 4 ⮞

Signals of Growth

During the past quarter of a century a steadily rising corps of professional investigators, numerous laboratories and institutes, and large sums of money have been devoted to the study of child development, care and management. Despite all of these endeavors [and] the volumes of advice and statistics concerning children which have been placed at the disposal of modern parents, it is readily admitted that men and women today approach the business of parenthood much less confidently than did parents a generation or two ago. The standard props, the familiar maxims which served our grandparents so well, have been knocked askew. Besides, even if mistakes of management were made in those days, they were not taken too seriously, for so great was the faith of our grandparents in the subtle powers of development that every child was expected to "outgrow" his undesirable traits as he merged into adulthood. Some did; some didn't.

In contrast, the modern parent is made to believe that the adult is to a large extent the product of parental treatment, that the molding of character and personality is cast during the early years, and the designer of those molds is the present parent. On the one hand, parents are confronted with considerable discourse about the importance of their roles, on the other, they face assertions by the same experts that the child is an individual whose inherent normal course of development should be allowed to unfold without undue interference. So emphatic have been these assertions that many parents have been led to believe that each child is a law unto himself; that there are no fundamental guiding principles which can be followed with any of the confidence that the familiar maxims gave to their forefathers.

It is true that each child is an individual, distinctly different from any other child. It is also true that the child is a growing, therefore a constantly changing, organism. For this reason a particular precept which might work at one time and in one situation might be hopelessly ineffectual in another, even for the same child. Fortunately, the days of dependable maxims are gone; irre-

Reprinted from *Child Study*, Fall 1940, 18:8–10, 31 by permission from Mitzi Wertheim.

trievably gone. In the light of our present knowledge it is admittedly difficult, if not totally impossible, to formulate rules as specific as "Spare the rod
and spoil the child" which could be applied, like a recipe, in the management
of any individual child of any age, and in any state of conflict with the parents.

Recognizing that no simple precept could apply to the management of all
children of all ages, much of the early work in child development was undertaken with the hope of providing some standards against which the individual child might be compared. Large numbers of children of all ages were
studied in order to determine the "average age" at which particular achievements should be made. Parents have been told to appraise the achievements
of their child by comparing him with this hypothetical "average child." The
natural consequence of such advice has been to create an undue interest in
special times at which the child might be expected to be capable of specific
achievements. These standardized tests have had the same effect appraising
behavior development that the height and weight tables had on physical
growth. The extent to which this emphasis upon chronological age has per-
colated to the popular mind is well demonstrated by sales clerks in toy departments. When one starts to make a purchase the clerk is likely to inquire,
"How old is the child?" If you answer 18 months she replies, "Oh, you
mustn't buy that; that's for a two-year-old."

The vanguard of those who work with children realize, however, that children cannot be fitted into an average pattern, and that chronological age, or
even mental age as measured by the standard tests provides no adequate criteria by which parents can guide their methods of management and education to suit the development of a particular child. The age range within
which any one activity may develop is too wide to justify the use of chronological or mental age as a reliable basis for altering educational guidance
procedures. So we find the old familiar maxims obsolete, and the newer
standardized tests proving not as fruitful as was hoped.

Does this mean that because children are individuals, changing from
month to month or even day to day, that they defy the formulation of general principles which can be of practical benefit in their education and upbringing? Or is it possible that the situation is not so chaotic as it seems; that
such principles do exist, but that the investigators have been too blind to see
them?

Theoretically there is every reason to believe that the latter possibility represents the true situation. A highly individualistic, haphazard course is not
the way of development in other types of life. Why should it be so with children? The farmer does not merely stand sympathetically by while nature
takes its course with the seeds he has planted in the earth. He has learned just
when and how much fertilizer to apply in order to obtain the best results.

Is it not conceivable that in course of time investigators of child development may be able to determine both the kind of stimulation required for maximum development and the opportune period for its application as successfully as the agriculturist has learned to modify the environment of the plants? Already there is a glimmer of dawn on the horizon. Since growth is fundamental to childhood, the obvious step is to ascertain *how* growth in behavior takes place. It is not sufficient to admit its existence or determine *when* it occurs. It is almost safe to predict that if the investigators of tomorrow will devote as much time and energy to studying the process of development as has been given in the past to the accumulation of facts and details about children, the parents of the new day may not have to search in the fog for guiding principles. Once the principles of development are revealed, methods of guidance and education can be modeled accordingly.

Already it is recognized that the methods of dealing with a particular child must change as the child develops. The infant who is just beginning to develop simple associations receives smiling approbation when he reaches for the car's ignition key, tries to push the starter button or blow the horn. Such antics show that he is beginning to recognize things in the world about him. His explorations are, therefore, met with approval. But when he is older he receives only frowns and scolding for the same behavior and is admonished "never to touch the dashboard." Such alterations in the treatment of the same child for the some act are as they should be, for he really is not the same child; growth has altered him considerably.

While it is generally agreed that methods of management should be altered in accordance with the child's growth and development, the primary problem is to recognize these growth changes when they occur. It has been stated that one cannot rely upon chronological or mental age as criteria. What then are the signals of development?

Studies of the development of infants have given use to the proposition that the signals for change are expressed in the behavior of each child; and that these signals were sufficiently homogeneous in their expression to be ascertained by the observant parent. For example, there is an opportune time to introduce a training program which will aid the child in the achievement of bladder control. By certain little ways of behaving the infant gives the signal that his nervous system is ready to undertake the building of simple associations. When the child gives some indication that he is aware of both the act of initiating and the result, he is then ready to respond to a training program. The behavior signals may be expressed in a variety of ways: he listens to the sound; he looks to the source; he assumes a quiet mien or has a glint in the eye immediately before the act; he plays in the puddles, or in some other fashion indicates that he is aware of the experience. Training begun before the maturation of these higher brain centers, as reflected in such behavior as was just mentioned, is relatively futile. Moreover, even if training is begun at the proper time, development in other aspects may temporarily interfere with his efficiency. As the child's ability to discriminate and generalize begins to develop, there may be a loss in the specific associations formerly held. For example, the association of voiding in one particular vessel may be temporarily diminished as he is forced to broaden his experience and discriminate between various kinds of vessels and suitable places. During this period the number of successful responses, even in the old familiar situation, is often reduced.

As the ability to discriminate increases, the child becomes more dependable in indicating his urges by gestures or other small signs. But he can't go on depending upon gesture language. When a growth spurt occurs in some other part of his development, diminished proficiency may temporarily occur in an activity which had formerly functioned satisfactorily. For example, as his speech begins to develop, the child may lose some of his skill in gesturing as a means of indicating his urge to void. While this onset of speech may check his ability at gesture-language, his speech is, in the beginning, too immature to serve adequately as a substitute way of communicating his needs. Hence another period of wet pants. Sometimes a child acquires a word for voiding, but uses it only after the act is begun; then you can tell that the newly developing speech centers of the brain are becoming connected with those centers which control urination. For a time they do not function simultaneously, so he cannot make his wants known early enough. After a while the two functions become coordinated, and the child can express his needs in words before urinating.

Another regression in bladder control may occur even after the child has for some time been anticipating and adequately voicing his urges. As the horizon of his interests widens and he shows concern with things beyond the immediate present, and both in his language and play he displays a tendency to fantasy, then we may expect another regression in the achievement of bladder control. At this time, the child is too occupied with these new experiences to take notice of the old familiar bladder sensation. Later when he is seen holding the legs tightly together, squirming, but at the same time continuing with his play, then one can know that the centers of his nervous system which are concerned with play are beginning to coordinate with those concerned with such functions as voiding. Soon he will be able to manage both circumstances satisfactorily. He will be able to drop his play temporarily, take care of his physiological urges, and return to pick up the play activity where he left it. When lapses in the achievement of bladder control occur, the observant parent should think not only of the success in this control, but should look for signs of rapid development in other aspects of the child's behavior. By learning to detect these signs of development in other activities the parent will understand and tolerate lapses in such achievements as bladder control.

Behavior signals can be useful guides in altering the feeding program of the infant. When the young infant extends his arms to reach for objects in his range of vision, he should be allowed to grasp the bottle or spoon and help in carrying it to his mouth. In this way the newly acquired grasping ability is incorporated into the act of feeding. Little by little he gains proficiency in conveying food from a dish to his mouth. But alas! Just as the mother is beginning to feel proud of this accomplishment, the child may start throwing the spoon about, spattering food all over the place, and in general making a messy nuisance of himself: Close observation of his mode of prehension (grasping) at this time may show that he is beginning to pick up crumbs, or other very small particles between the thumb and the tips of his fingers. This change in method of prehension is the signal to stop the pureed vegetables and ground meat, to give the child at least a part of his meal in small particles so that he can exercise this newly developed ability to pick them up between his fingers. Certainly this is not the time to be concerned with table manners. Later interest in manipulating spoons will recur. At this time the child may tap the food or plate with the spoon, or in other ways indicate that he is ready to start learning the act of dipping the food on to the spoon. In this way the child acquires proficiency in self-feeding. Even though he becomes quite accomplished during the first half of the second year, there will be times months later when he will ask to be fed, or dawdle over his food. This change in behavior is aggravating unless one can realize that the phase parallels rapid development in imagination, fantasy and interests outside the routine things of his life.

With this expanding interest, as reflected in topics of conversation, the child is also beginning to develop an appreciation of his own independence. These regressions are expressions of a transition from the dependence of an infant to the independence of childhood. At times he is reluctant to give up his infantile dependence upon the parent. It is probably wiser to give aid on such occasions than to remind him sarcastically that he was able to feed himself much better when he was only a baby. For example, a three-year-old boy visiting a fifteen-month-old baby girl, who fed herself admirably, was shamed by his parents because he asked to be fed. That selfsame baby girl, who is now three years old, often wants be fed, not because she cannot do it herself adequately but because her mind is so given to excursions on the events of the day. At fifteen months she gave attention to the act of feeding because she was not capable of experiencing the interfering diversions. Later, when these two functions will become better integrated, she will be able to eat and converse about other things at the same time.

In watching for the behavior signals of development it must always be remembered that the child has an indomitable urge to exercise his newly developing function, but that once he has acquired an appreciable degree of proficiency he grows indifferent toward it. Recognition of this fact may avoid many storms and tantrums. The child who is just learning to identify the parts of his clothes and just where they go, will tax the busy adult's patience because he insists on dressing himself, although he cannot do it adequately; and the same child, once he is capable of dressing himself, will dawdle or demand help in his dressing. And the parent's patience snaps because the child refuses to dress himself when he so obviously can do so.

These illustrations are given merely to indicate what is meant by watching for signals in the child's behavior which can be used as a means of selecting the best guidance in providing for this environment. Modern investigators in the field of child development cannot provide specific prescriptions comparable to the maxims of old, but they can be helpful in determining the types and meaning of behavior signals which may be expected in the child's essential activities. It is reasonable to assume that the child shows as definite signs of readiness to begin to learn how to read as he does for toilet training.

The obligation of future research workers in child development is to point the way so that parents and teachers may be better enabled to detect and interpret these behavior signals. But the big job still rests with the parents. They must learn to observe. The ability to observe is not a God-given trait; it expands and develops with experience, whether one is observing the behavior of children, a slide under the microscope, or birds in their flight. A peculiar rustle among the leaves will tell the observant person it's the scratching of a Towhee, while to another it is merely a noise in the leaves, and the still less observant person will not even hear it. A few hints about form and rhythm will be helpful to the beginner in art appreciation, but even the best critic cannot tell the student just what to see in a great painting. The first requisite in becoming a good parent is to learn the art of observing children. Learning to observe and respond to the minute signals in the behavior of the infant and young child is like learning to listen to classical music, to detect the signs of pneumonia in an X-ray film, or to see more in the night sky than a conglomeration of stars. One must make a deliberate effort to learn to observe and interpret the signals of growth in behavior. When that ability has been acquired, modern parents will have confidence transcending anything provided by the old adages. Fortunately, this is an ability which expands with experience; and with enriched experience, confidence in one's self as a parent is enhanced. Self-confidence is one of the primary requisites of sound parenthood. To gain parental confidence in this way is more difficult than to follow the rule-of-thumb technique—but it also is a lot more fun.

❧ 5 ❧

The Experimental Twins

Some eight years ago the story of a fourteen-month-old baby scooting around on roller skates and swimming with his face underwater caught the eye of the newspaper world. The story literally went around the earth as evidenced by clippings which have come in from China, India, Russia, Chile and other remote places. The baby was one of a pair of twins who through nearly a decade have retained their reputation as "the experimental twins." Since the original release the story has been revived annually on the occasion of the twins' birthdays. During this time all sorts of persons, except the investigator, have undertaken to interpret the study and its significance to the public. In the public mind the study has been accepted as a kind of race to see which of the boys will make his mark in the world. People like to take sides; some root for Jimmy and some for Johnny. Even grotesque ideas about the purpose of the study are not uncommon and the belief is frequently expressed that the training of a baby should in some way improve his general intelligence or mental endowment. Because of these false notions and because the writer can scarcely get through a peaceful dinner party without having someone ask, "What is the difference between the twins now?" this attempt is being made to explain the purpose and outcome of an investigation which should, perhaps, never have gone beyond the boundaries of scientific publications.

Let us at the outset clear up some of the common mistaken ideas about the experiment. Was the purpose of the study to prove whether heredity or environment is stronger in determining the outcome of the individual? The answer is no. Was it to make a genius or a superior child out of an ordinary or dull one? The answer is no. Was one baby kept at the hospital all the time in the care of a trained child psychologists while the other remained at home in the custody of the mother? Again the answer is no.

The study was begun in 1932 at the time the children were born. An earlier study on the training of infants in motor behavior had come to the con-

Unpublished essay, 1942. Included here by permission of Mitzi Wertheim.

clusion that the performances of infants could not be improved through practice or exercise; that during babyhood one must merely wait patiently until the cells of the central nervous system ripen or mature and are ready to go into action. These conclusions were, apparently, contradictory to the famous edicts voiced by the Behaviorist, John B. Watson, who achieved popularity with parents when he claimed that, given a dozen healthy infants, well formed, and his own specified world to bring them up in, he could make of them whatever he chose—doctor, lawyer, artist, merchant, chief or even beggar man and thief. The Behaviorist's doctrine was a hopeful one. We, as parents, like to feel when we first hold our little creations of human flesh that we are not only custodians, but molders of their destiny. The theory that infants do not respond to training presented a doleful outlook. Parents like to feel that they have a job to do more than sitting by while nature takes its course. Moreover, we do know that practice and training are not without influence in the development of older children. While the old adage—practice makes perfect—may not be entirely true, no one would deny that practice usually improves performance. If it is not important during infancy, when does it begin to count?

The twin experiment was begun with the thought that by exposing a baby from the very first to a definite amount of muscular exercise it would be possible to determine the time in the life of the individual—somewhere during the change from infant to childhood—when improvement of performance could be brought about through practice or exercise. We, therefore, selected Johnny, because he appeared to be more frail at the time of birth, as the baby to receive the special exercise. When he was twenty-one days old, we began to exercise him daily in all those little motor movements of which the young infant is capable. During the early months, he was given four exercise periods a day, each period requiring about ten to fifteen minutes. Both babies were, from the very beginning, brought to the laboratory about nine o'clock in the morning and were returned to the home of their parents about five in the afternoon. They were always at home on Saturdays, Sundays, and holidays. Jimmy, as well as Johnny, was brought to the laboratory daily because he played a different but no less significant role in the experiment. Jimmy's routine was more conventional. Young parents who were delving into current books on child care during the twenties and early thirties, will recall that it was a vogue at the time to put a baby on a regular schedule, regular according to the clock, and to guard against too much handling or over-stimulation. It was thought that if the baby were over-stimulated he would become a nervous or neurotic child, or that his delicate digestive system might be disturbed. So during the time the twins were at the laboratory, Jimmy was kept on a schedule. Except for the handling occasioned by his routine care—feeding, bathing, etc.,—he remained in his crib in the nursery.

His chances for activity during the laboratory hours were relatively restricted, especially as he grew older.

The type of exercise given Johnny during the early months was simple. He was encouraged to make crawling and stepping movements; he was held by his feet in an inverted position to stimulate activity of the spinal muscles; his grasping and other reflexes were stimulated daily. As he grew older and became capable of more complex behavior, his practice schedule was expanded. For example, when he began to reach for toys held in front of him, he was given daily practice in that function. When he began to creep on his hands and knees, he was given daily exercise in creeping up and down inclines of varying steepness. As soon as he could sit on a stool and turn himself around, he was exposed to daily practice in getting off pedestals of various heights. When he began to stand on his two feet and take a few steps he was placed on roller skates, not with the thought of teaching him to roller skate, but primarily to make the job of balancing a little more difficult for him and to stimulate activity of certain muscle groups and nerve centers which were just beginning to work together. Since the spontaneous movements of new-born babies simulate swimming movements, Johnny was supported in a tank of water to stimulate the use of muscles involved in swimming. When he was a little older, it was observed that he, like most toddlers, enjoyed pushing furniture around the room. So the situation was set for him to push pedestals of different heights around and arrange them in order to climb up and get a piece of banana or cracker suspended from the ceiling.

The outcome of such practice in motor activities is well known. A baby of less than a year was scaling slopes of unbelievable steepness. At fifteen months he was doing a credible job of roller skating and in less time he was swimming twelve to fifteen feet with his face under water. These are the achievements which attracted popular notice. Popular reports neglected to mention that while he was learning to roller skate, he had at the same time been pushed up and down the corridor daily on an ordinary tricycle. For seven long months there was hardly a glimmer of improvement in his performance in this apparently simpler act. It is regrettable that the occasions in which Johnny failed to show appreciable improvement were not given notoriety along with his achievements. Any interpretation of the achievements which does not take into account the failures is without foundation and meaning.

Then what does the outcome mean? The answer is clear and simple. Like most controversial issues, the truth is somewhere between the extreme claims. The development of children is not merely a matter of waiting for the nerves to grow; nor can one take any child and make anything they want to out of him merely by controlling the environment in which he shall live. In training young children one must wait until the nervous system matures, until it achieves a state of readiness to perform. But the child does not achieve this state of readiness in all things at the same time, nor do all children achieve it at the same age. However, once the child has attained a state of nerve and muscle preparedness, he will profit much by an opportunity to exercise those muscle groups which have just begun to work together. By putting Johnny on roller skates just at the time his balancing and locomotor systems were beginning to work together, we were able to graft the roller skating performance on the act of walking. Johnny not only exceeded the accomplishment of Jimmy in this activity, he exceeded our expectations of any baby of comparable age. It is probably safe to say that the toddler will learn to roller skate with less expenditure of energy on his own part than will the older child. Although the baby has a less mature nervous system to work with, he has a less difficult mechanical task to accomplish.

Obviously it is possible through training to advance the achievements of an infant beyond that which is normally expected. On the other hand, it is just as important not to introduce practice too early; that is, before the child is ready for it. It is just as foolhardy for one to start teaching the average two-year-old to read as it was for me to try to teach Johnny at the age of eleven months to ride a tricycle. There are opportune times in the life of an individual when he can most economically acquire any particular activity. Our job as parents and teachers is to learn to detect when those precious moments arrive. One must observe and watch for little signs in the child's behavior to indicate when he is ready for training in any particular type of activity. Neither chronological age nor mental age alone is adequate criteria because children vary in their way of developing. Of course, it does not mean that if training is withheld just when the child is ready he will fail to acquire the activity. It does mean that he will expend greater nervous energy in doing so later. Johnny's readiness for tricycling developed when he was about twenty months old, but in the meantime he had acquired a bored and sophisticated attitude toward the activity so that he did not meet it with the interest which it would have had for him had the tricycling been withheld until his neural and muscular systems were capable of handling the situation.

During the first twenty-two months of the twins' lives, while Johnny was getting his daily work-out, Jimmy had to rely more upon his own resources for entertainment. As stated above, he was kept more or less isolated in the clinic nursery except for the handling necessitated by his routine care. Also, at intervals, he was tested in the activities in which Johnny was receiving daily practice. In the nursery Jimmy had plenty of opportunity to hear other children and adults talking and also to chatter away for his own amusement. Restricted in his motor activity, Jimmy played in his crib, chattering a blue streak as soon as he became old enough to vocalize. Remember, Johnny's practice performances were essentially motor and not linguistic. When the twins were about eight months old, distinct differences in their motor performances became apparent. This difference became progressively greater during the second year of their lives. While Johnny was gleefully scaling steep inclines and jumping off high pedestals, Jimmy would not attempt to climb even the mildest slopes. Contrast in their general attitudes was even more striking. At the laboratory during those first two years, Johnny was the epitome of happy babyhood. As the experimental period wore on, Jimmy became more and more tense and during the latter part of the second year was subject to violent temper tantrums. These effects of frustration increased as Jimmy grew older and his restrictions became more of a handicap.

It was primarily because of these behavior symptoms in Jimmy that the experiment was not carried beyond their second year. It was for the same reason that during the last three months of their daily visits to the laboratory Jimmy was given intensive daily exercise in all those activities in which Johnny had shown special accomplishment. We did not feel justified in turning a cantankerous child over to an already heavily burdened mother for daily supervision without in some way trying to recondition his attitude. Also, being humane, we wanted to pay our debt to Jimmy for his long period of restricted activity. During the period of special training, Jimmy made rapid advancement in motor performances. His tendency to chatter did often interfere with undivided attention to the task of the moment but nevertheless, he showed great improvement in most activities. He never achieved Johnny's proficiency but he did in many instances approximate it. During this time he became much more cooperative and cheerful when in the laboratory. We can only guess whether more permanent effects would have been imposed had his period of isolation continued longer.

Special practice of both children was discontinued when they were about two years old. Since then they have lived at home during the day time as well as at night. They have, however, returned to the laboratory at intervals for a check-up on their motor performances. Next April the children will be ten. What has been the effect of their different experiences during the first two years upon their subsequent development?

The question is natural but the answer is not simple. It is not simple because the appraisal must be made in terms of specific activities. It is not simple because life for the children became more complex, and a multitude of influences beyond experimental control contributed to their course of growth. However, it is possible to evaluate their subsequent performances in those specific activities in which they were given special practice. In certain acts, such as tricycling, neither Johnny nor Jimmy exhibited loss of skill even after practice was terminated. In others, such as skating and climbing inclines, there was a marked change. In some instances the loss in skill could be attributed partially to the changes brought about by the child's growth in physical dimensions. Loss of skill may not be merely a matter of forgetting. The act of skating, for example, is a different problem for the six year old than it is for the toddler. It is different because the child is different; the legs are longer and the center of gravity is lower. It is like giving a child two different sets of tools and telling him to make exactly the same thing.

When all activities are considered, it is safe to say that insofar as quality of performance goes, Johnny has consistently exhibited greater motor coordination and an ability to handle his muscles with ease and grace. Jimmy has rather consistently exhibited greater linguistic facility. Johnny has also shown greater fortitude in performing difficult feats, such as jumping. This superiority in motor skill and courage in some instances operated as a handicap. For a long time Johnny would arrange the pedestals in a careless, haphazard fashion since he knew full well he could easily bridge the gaps between them. Jimmy, on the other hand, would arrange them in a neat little stairway, though he would hesitate to climb up on the tallest one after he had made the orderly arrangement. Both children have worked out their own salvation, their own systems of compensations. Jimmy can distract from his less skillful motor performances by a long line of engaging chatter. In appraising his motor skill even the experimenter has at times been hoodwinked by his amusing antics until the silent film was processed and his motor performance could be viewed stripped of the lingo.

Independent judgments of the children's emotional and personality makeup, as determined through interviews with a child psychiatrist and by their reactions to a standard personality test, agree that Johnny is more mature and better organized in his emotional reactions to the world in which he lives; that Jimmy responds more like a typical child of his years, that he is an extrovert who lives for the moment while Johnny is more contemplative, more stoical, and better able to face disagreeable situations.

Just what kind of adult each boy will become is still to a large extent conjecture, but it is an issue which has no real bearing upon the purpose of the experiment. It can be said that motor exercise during infancy contributes to better coordination of muscular activity in later childhood. To the investigator the study means that there are opportune times in the life of a child when he can most advantageously expand his development in any given direction. The problem for parents and teachers is to learn to detect those critical moments and to give ample opportunity for the child to exercise each function when the time is ripe. Between six months and two years the child is developing rapidly in the ability to use and control his muscles. It is safe to say that if a baby is given a chance to exercise his muscles during this time, he will gain greater smoothness or coordination in muscle function than he otherwise would have had. Of course, in vicarious ways such as rough-housing with father, many babies get such stimulation even though it is not deliberately or systematically imposed.

Learning to observe and detect just when and how much to stimulate a child in order to advance his own potentialities is the art of parenthood. Fortunately it is an art which grows with experience. The more one observes (with a questioning mind) the more he will learn about his child. The better he knows his child, the greater will be his enjoyment of parenthood. Fortunately, the bugaboo of overstimulation no longer hinders us. Parents are now advised to enjoy their children and to demonstrate affection for them. We now realize it is not a question as to how much stimulation is good for the baby but rather the manner in which stimulation is presented.

⊷ 6 ⊷

Infant Motor Development: A Study of the Effect of Special Exercises

One thing which all civilized people share is the desire to construct methods of education which will allow every individual to develop both physically and mentally to the extent of his potentialities. It is also true that our basic convictions about the nature of growth and development will influence our educational theories and methods.

In America, before and during the 1930s, we were drawn between two points of view—that of the "Environmentalist" on the one hand, and of the "Maturationist" on the other. The "Environmentalists" claimed that the individual was molded by the external forces exerted upon the pliable nature of the infant and young child. The "Maturationists" contended that development during infancy was dependent upon the ripening processes inherent in the nerve tissue and was, therefore unaffected by practice or exercise. It was with reference to this controversy that the study being reported was undertaken in the early 1930s.

This study was only a small part of a larger investigation involving more than a hundred babies. The purpose of this broader study, which was in collaboration with the late Dr. Frederick Tilney, was to ascertain the correlation between overt behavior and cellular structures of the brain. At that time it was customary in the field of neuro-psychiatry to regard mental disorders as either "organic" or "functional." If organic, the ailment could be identified with a specific lesion in the nerve tissue; if functional, the etiology was presumably to be found in the personal experiences of the patient.

It was within the theoretical framework of these controversies that we set out to learn if motor development of infants could be expanded by practice or exercise. The subjects for this special study were twin boys—Johnny and Jimmy. At birth Jimmy appeared to be better endowed in muscle tone and

Lecture to an International Conference on Sports Medicine, Moscow, 1958. Included here by permission of Mitzi Wertheim.

117

strength; therefore, Johnny, the weaker twin, was selected for special daily exercise in motor activities of which he was somewhat capable. Such practice was continued throughout this first two years. Although the daily exercise of Johnny was discontinued when the twins were two years old, both boys were brought back to the laboratory periodically for check-ups until they were nine years old. Then when they were 22 years old we were able to obtain a short film of their general motor coordination.

The outcome of the practice period during infancy may be summarized briefly as follows: During the early months Jimmy, who was not getting special exercise, advanced beyond his brother in the "emergence" of basic motor activities, such as sitting, reaching, creeping, walking, etc. But, after the first nine months Johnny began to outstrip not only his twin but also the other one hundred infants under observation in such activities as swimming, jumping, climbing, roller-skating, etc.

What did this mean in terms of our knowledge of the nervous system? By means of histological studies of the newborn brain we had reason to believe that the cerebral cortex does not function at birth even though the six layers of cells are present. Therefore, it could safely be assumed that all activities of the newborn infant are under the dominance of sub-cortical or mid-brain centers. Additional histological evidence indicated that the first area of the cerebral cortex to show cellular differentiation was the motor area. Correspondingly, the type of behavior growing most rapidly during the first fifteen months is motor activity. Another significant observation was that an element of "awareness" enters into the baby's motor activity about the ninth month. It was this element of "awareness" which prompted the conjecture that the cerebral cortex was now functioning together with the mid-brain and other lower centers. It was this element of awareness which led to the conclusion that "motor performances can be expanded during infancy beyond that normally expected if exercise is provided at the time the cerebral cortex comes into play" (McGraw 1935:311).

The next question is: Does such exercise during infancy have any permanent effects? The follow-up records on these two children during the first nine years together with the films taken when they were 22 years would indicate that it does. This assertion can be better demonstrated by the accompanying films than by my statements. [McGraw indicated that the first part of the film showed the advancement of Jimmy, the untrained twin, in basic motor activities. The second part illustrated Johnny's motor performances during the period of practice while the third demonstrated the motor coordination of both boys at the age of 22 years.]

Before turning to the film, I would like to leave with you one thought, not as a conclusion, but as a challenge for future researchers and educators. Obviously the motor activities of infants can be expanded if exercise is provided at the time the neural centers of the brain have reached a state of maturational readiness. If this is true of motor development, is it not reasonable to assume that it might be true in other areas of growth and learning? The problem is to establish the opportune time, or state of readiness, for each individual and each function. In this connection the outlook is more promising than it was in the 1930s. We are no longer dependent upon histological studies of dead tissue for our knowledge of the nervous system. Electronic devices, e.g., the electroencephalogram, and biochemical techniques are opening new doors for a better understanding of nerve impulses in action. What we need most urgently now is inter-professional collaboration between the neurophysiologists and educators and others, so that our educational theories and methods can be brought into harmony with our advancing knowledge of the living brain.

7

The Problem of Using Secondary Sources: Elkind's Blunders

[Editors' note: McGraw wrote this essay to correct what she considered to be a misrepresentation of her research. David Elkind contended in his 1986 essay that parents and educators held the misguided belief that pre-school children could benefit from formal instruction in reading and other academic skills. He argued that there was no evidence that such efforts accelerated intellectual development and much evidence to suggest that early education was harmful to children's emotional development. Elkind cited McGraw's research in early motor development to defend his assertion by suggesting that Johnny's early motor training failed to confer any advantage in learning. He also asserted that a psychiatrist who examined Johnny and Jimmy (see Dennis, Introduction to Part 2) told him that Johnny had become more insecure following the experiment. McGraw took exception to Elkind's account of her experiment because she did not employ academic techniques of instruction—techniques she strongly criticized—and demonstrated that early motor stimulation increases the motivation to learn. McGraw also had good reason to be dismayed by Elkind's reference to the psychiatric evaluation of the twins. Dr. Langford, the psychiatrist who evaluated the twins, actually observed that Johnny felt less secure at home (see Chapter 4) because Jimmy was given more attention than Johnny at home. Importantly, Langford concluded that he did not believe "that undue pressure for success in various accomplishments was a factor in this as the boy enjoyed his work."]

Recently a colleague brought me a reprint of the article by David Elkind which was published in the May, 1986 [Phi Delta] *Kappan*. Much to my chagrin the major point of the article was exemplified through a misrepresentation of my research efforts, results, and publications.

Unpublished essay, July 23, 1986. Included here by permission of Mitzi Wertheim.

To start, I am and have been, in complete agreement with the position Elkind supports, which is, that the typical classroom teaching design for school age children is inadequate for the instruction of younger children.

Let me state that I found the article to be engrossing, stimulating, and hopeful. It lamented the lack of an organized systematic method for the education of young children. In doing that, it fits well with other contributions to the media which have already begun to arouse groups of the populace over the dilemma.

Early childhood education is at present a subject of intense controversy. Some contend that the education of younger children should be assigned to the public school system. Others caution that the processes of learning and development are so different during the early years from those of school age children that traditional classroom instruction is unsuitable for children under 5 or 6 years. Elkind identifies himself with the latter position and he does so admirably. He is a gifted writer, but more significantly, he has the skill to detect the impact of innumerable intertwining forces that may affect any generalization about the teaching and learning of the young. As a reader I was impressed and fascinated with the article until page 635 where he proposed an anecdotal illustration pointing up the possible damage a learner may experience from formal instruction.

Although referring to the study of Johnny and Jimmy that I conducted during the thirties, the anecdote has nothing to do with it. Instead it deals exclusively with an interview Elkind said he had with a well-known psychiatrist "several years ago." Let me emphasize that the anecdote in no way diminished my admiration for the substance of Elkind's article which was concerned with the current arousal and controversies about the methods of early childhood teaching.

While studying Johnny and Jimmy, along with a hundred or more other children during the 1930s, I gained the impression that the first self-motivated manifestation was curiosity followed by the urge to explore or investigate. Perhaps current educators should consider that idea as a promising basis for formulating new techniques for teaching infants and pre-schoolers. Of course I was puzzled why Elkind had chosen to thrust that particular anecdote into the midst of an otherwise super article.

Now for my rebuttal:

1. The most serious faulty assumption of the anecdotal presentation was that Johnny was "taught" or shown how to perform in a particular situation. Nothing could be farther from the truth. We were not then interested in teaching techniques. We merely arranged the equipment, placed a lure (any lure), and observed the infant's or toddler's performance, recording it in written notes and on 16 mm. film. The changes, as recorded from time to time, revealed oscillations, rhythms, and fluctuations—characteristics of the processes of growth and development, or, if you wish, self-learning.

2. The anecdote assumes that the program of investigation followed a rigid schedule: Johnny receiving daily practice with the laboratory equipment during the first two years, and Jimmy receiving none. That is a gross misunderstanding, a blatant error of facts. Johnny did receive daily practice, but Jimmy also received a work-out on the same equipment at regular scheduled intervals, and, just before we withdrew the daily visit to the laboratory, Jimmy was exposed to daily practice for one month. All of these details are reported in *Growth: A Study of Johnny and Jimmy* and on 16 mm. film (donated to the University of Wisconsin).

Less critical, but of concern:

3. The anecdote doesn't reveal the name or the psychoanalytical orientation of the "renowned psychiatrist."

4. It states that he saw the twins "several years" after the study was terminated. The study was terminated during 1940 (during 1941 this nation was involved in World War II). That meant that his meeting with the twins was about forty years before the psychiatrist's interview with Elkind.

5. The age of the boys at the time of the psychiatric interview is not indicated. Were they being seen by him because of some behavior abnormality for which they had sought expert advice or had he, out of personal interest, sought to examine them?

I was so bewildered by Elkind's insertion of that particular anecdote that I searched to understand his motivation for my own satisfaction. I speculated for a reasonable explanation; Elkind was, given the current educational turmoil, so negative to the idea of school systems adapting school age programs for the education of young children, that he thought a human illustration of its possible negative effect would strengthen his general argument. But it led to a misinterpretation of both the purpose and outcome of my extended study of Johnny and Jimmy during the 1930s.

Education of school children as well as of younger children could be improved with some basic reform. I have often contended that the original designers of the public school system made a great mistake when they decided to bundle children of the same chronological age together in a class and expect them all to learn the same specific subject at the same time. But we must remember that at the time the school system was designed the bulk of learning took place in the home and on the farm, not in the school. It was enough for the schools to teach the three r's—reading, writing, and arithmetic.

The newborn infant consists of a bundle of soft tissues and neural, muscular, and organic structures bounded by skin. It has energy for considerable activity, some of it random and others specific (reflexes). Can you imagine that anyone would suggest that a study of infant behavioral activity may be of use to us in developing our frame of reference for the improvement of education of growing school-agers? That is the proposal I introduce for consideration. We usually think of education as expanding knowledge. One thing I discovered during the thirties was that a few specific behavior patterns present at birth were set for decline within the first three or four months. So, theoretically, decline as well as emergence should be taken into account as part of the processes of development and learning. In other words, some things must decline to provide for the emergence of new performance. Examples are reflexes, such as the Moro, Darwinian, and swimming—they begin to decline soon after birth. Sneezing, coughing, etc., etc., are also reflexes, but so far as I know they stick with us throughout life after starting to function soon after birth.

During the 1930s I devoted myself to an intensive longitudinal study of the behavioral development of a number of infants—in addition to Johnny and Jimmy, beginning at birth and extending through the early years. Educators are now in a turmoil about public schooling. The answer to their problem is not, however, to merely extend the school day and add the younger children to the classroom. At least some community sources might offer new designs. As a teacher during the sixties I organized a laboratory to provide for my students an opportunity to work with the growing infant. It was based upon the design of my investigations of the thirties—except that each student worked with an assigned infant through the course. The letters I have received when those students marry and have their own babies indicate that they profit by that limited exposure to observing infant development throughout one academic year. Some elementary and high school students would profit if they could be given the opportunity to cultivate the art of observation, recognizing the signals of development and cultivating the skill of communicating with the prelinguistic child, along with the sensitivity to allow the infant to express and expand on her/his inner motivation. If a community obtains cooperation of the hospitals, the day-care centers, and the pre-school nurseries, experimentally it might be possible to educate interested elementary and high school students to experience the opportunity to consciously and sensitively comprehend the processes of behavioral growth and development. In time maybe the educators themselves will recognize that there are other areas beyond the "academic" for the preparation of adulthood. They shouldn't have to wait until they've borne their own child to theoretically prepare for parenthood.

PART THREE

Coghill, Neuroembryology, and the Principles of Development

Introduction: McGraw's Alternative to Gesell's Maturationist Theory

THOMAS C. DALTON

As suggested previously, McGraw's research was overshadowed by Gesell's work largely because most researchers in child development believed erroneously that her studies supported Gesell's maturationist theory. The essays and chapters from McGraw's books—*Growth* and *The Neuromuscular Maturation of the Human Infant*—that describe the premises and methods underpinning her work are included here to dispel this misconception. Significantly, most contemporary researchers have been generally unaware, until recently (see Bergenn, Dalton, and Lipsitt 1992) that Coghill's ideas and methods played a prominent and more instrumental role in McGraw's research than that of her rival, Gesell. Thelen and Adolf (1992) argue incorrectly that Gesell used Coghill's ideas to support his theory that neural development was genetically determined when, in fact, Coghill never held such a position. Consequently, the controversy over maturationism pivots on a clear understanding of Coghill's ideas and precisely how McGraw and Gesell employed them differently in their research. This analysis will also identify key differences in McGraw and Gesell's explanations of development, and review how their theories have fared in light of recent evidence about developmental processes.

COGHILL'S CENTRAL INFLUENCE

Perhaps the most distinctive and significant feature, setting McGraw and Gesell's infant studies apart from their contemporaries, was their common interest in drawing from embryology to illuminate the processes of behavioral development. Experimental embryologists such as Hans Spemann, Paul Weiss, Ross Harrison, George Coghill, and others contributed new discoveries in the 1920s and 1930s about the sequence of early growth and dif-

127

ferentiation processes that were highly suggestive as to potential factors shaping subsequent stages of development. Yet the influence that these and other pioneering developmental biologists have had in the fields of neuroethology and psychology remains poorly understood and unappreciated (Oppenheim 1992). George Coghill, a neuroanatomist, first identified neural growth mechanisms governing the behavior of the fetal salamander. He argued against the conventional wisdom that behavior is constructed from chain reflexes by contending that specific behaviors emerge through the progressive expansion of a previously integrated total pattern. This was based on Coghill's (1936:10) discovery that local segmental responses occur not as a result of stimulation but by virtue of being released through inhibition of an integrated response. Coghill's conception of integration was not put forward as a genetic principle of development, but as an alternative to the orthodox view that reflexes first emerge from undifferentiated mass reflex activity. He argued that fetal behavior exhibits a high degree of coordination (involving the alliance of movements within a unitary pattern) and that therefore, an integrated response represented the *normal* or characteristic feature of neuromuscular action (Coghill 1936:11–12).

Coghill's work (1929; 1933a) aroused substantial interest among psychologists, but few experimentalists possessed the background necessary to understand his concepts and apply his methods to the study of infant behavior. The controversy, stirred by the mistaken belief that Coghill was saying that development is governed solely by endogenous factors, and thus, genetically determined, diverted attention from the significant ramifications that Coghill's thesis posed for understanding the relation between learning and development, into a debate about heredity and environment (Oppenheim 1992; Bergenn, Dalton, and Lipsitt 1992). Although Coghill never took an explicit position on the maturation-learning debate (Oppenheim 1978) he nevertheless closely followed Gesell and McGraw's studies because of the potential implications for his theory.

Gesell and McGraw were both strongly influenced by Coghill's ideas and became personally acquainted with him in the early 1930s. Gesell struck up a correspondence with Coghill in 1926, exchanged reprints with him, and solicited his critical reactions to his theory of development. Coghill visited Gesell's laboratory at Yale on at least one occasion in October, 1933, combining it with his first visit with McGraw in her laboratory at Columbia University (Coghill 1933b). According to his biographer (Ames 1989:283), Gesell sought Coghill's support for his theory because Gesell believed it corroborated Coghill's findings. While Coghill (1934) concurred with Gesell's contention that "environmental factors 'inflect' but do not generate the progression of development," Coghill (1939) withheld final judgment on the validity of Gesell's penultimate theory of "reciprocal interweaving" to explain infant development.

McGraw first contacted Coghill in 1931, requesting a copy of his films on the salamander (Coghill 1931). McGraw was just getting her studies underway at Columbia, unlike Gesell, and could incorporate Coghill's suggestions into the experimental design. The Josiah Macy Jr. Foundation provided a grant to support Coghill's research at the Wistar Institute in Philadelphia that apparently included informal consultation with McGraw (Herrick 1949 and 1944). McGraw acknowledged years later Coghill's extensive involvement and influence in her research by saying: "Coghill visited my laboratory many, many, many times—sometimes with Tilney, sometimes not. We talked and exchanged ideas. It was he, John Dewey, and the babies that got me thinking of process, not end result, or achievement" (McGraw 1979).

McGraw detected some important anomalies in the behavior exhibited by infants in neonatal development that led her boldly to reject the "chain-reflex" theory of development in favor of Coghill's view that integration precedes processes of individuation, highlighted in her essay: *The Functions of Reflexes in the Behavior Development of Infants*. For example, McGraw observed that neonatal infants could grasp a rod tightly enough to be suspended in the absence of adequate muscular development in the shoulders and upper arms. She believed this suggested that embryologists such as Coghill were correct in concluding that limb buds are differentiated first and innervated prior to these other segments, indicating that early neural growth anticipates the acquisition of function (McGraw 1932; 1933). The question of whether spinal motor axons make contact with limb muscles prior to innervation, or only after motoneurons are specified remains an important but unresolved issue among developmental biologists (Jacobson 1992:470–474). In addition, McGraw speculated that reaction patterns that exhibit a characteristic "total body response" such as the diffuse Moro reflexive reaction to surprise, were not simply clinical signs of waning, non-functional phyletic traits whose persistence is associated with abnormality, but constituted precursors or forerunners to behavior subject to cortical control.

McGraw's suspicions about the limitations of clinical measures of normal infant development were reinforced most pointedly by her discovery that some newborns could perform behavioral feats considered impossible until a later age, such as rolling from a prone to supine position. Physicians discounted such movements, attributing them to excessive tonicity or muscle spasms, refusing to consider the possibility, according to McGraw, that such advanced behavior may be indicative of a greater degree of neural development (Chaney and McGraw 1932:50). McGraw recalled her frustration in applying standard measures in the following anecdote:

> As I watched the newborns in the delivery room in the obstetrics nursery, I questioned the adequacy of standardized test items to reveal the quality of an activ-

ity. For example, the Gesell developmental scale rated turning from back to stomach at 4 months. But occasionally, not too often, I saw newborns in the delivery room or the nursery manage somehow to get from back to stomach. ... It was not just the newborn getting from back to stomach which ignited my doubts about static objective measures. I also noted that if a newborn is left alone in a crib he may migrate from one end to another by flexing the trunk from left to right and moving his arms and legs somewhat rhythmically. Should that be scored crawling or creeping on standardized scales? When is an achievement an end result of an ever changing function? Is the end result only in the experimenter's judgment? How to appraise the quality of an activity in an ever-changing organization? These were the questions confusing me in the early months of the laboratory (McGraw 1980:39–40).

In retrospect, it is difficult to understand why psychologists continue to hold that Gesell and McGraw conducted essentially equivalent studies of development and learning processes despite significant differences in methodology and findings. For example, Gesell (1934a:38) began his observational studies at 4 weeks rather than from birth. Gesell (1934a:315) did not introduce special training until the experimental twin was 11-1/2 months (and after 14 months for the control twin) compared to 20 days for Johnny and Jimmy (see McGraw 1935:40). (Jimmy was not afforded special stimulation but he was tested in the same activities at the same interval as Johnny.) Gesell (1934a:40–41) also used repetitive training exercises, administered tests that heavily emphasized vision, perception, and memory, and conducted no follow-up studies. Gesell's (1934a:316–317) findings that training neither alters the sequence nor accelerates the processes of skill acquisition, and that delayed practice results in greater gains should not be surprising given these differences in methodology. The twins in Gesell's studies had already attained some degree of sensori-motor development at the onset of their respective experimental training and testing programs. Nevertheless, many aspects of McGraw and Gesell's experimental investigations remain inconclusive regarding the effects of stimulation or restriction on learning. This has prompted Razel (1985; 1988) to urge the establishment of new experimental co-twin studies that can be designed specifically to address issues left unresolved by McGraw, Gesell, and Wayne Dennis' pioneering research. This suggestion merits serious attention because it would offer the opportunity to incorporate the latest knowledge and techniques of developmental biologists, in fathoming the connections between neural growth processes and developmental behavior.

McGraw's Experimental Methods

McGraw confronted her doubts about age-based norms by adopting methods that would more clearly identify how growth processes account for the transient forms that behavior assumes in the course of development. There are important similarities and differences in the perspectives that McGraw and Gesell adopted about growth and development, long obscured by the maturation controversy, that help explain their contrasting conclusions about the effects of early stimulation on learning and maturation processes. Gesell (1934a:294) and McGraw (1935) concurred that growth provided the impulse towards the organization and optimum realization of functional capabilities. However, Gesell concluded that early training does not accelerate growth, nor does it alter the sequential acquisition of functionally specific skills in development (Gesell 1934a:308). Although acknowledging that growth, development, and the differentiation of structure and function are not easily distinguished, Gesell (1934a:293) preferred to define growth primarily as augmentation of function. This was not unwarranted since no common unit existed to measure growth as a process rather than an outcome. Instead, Gesell gauged developmental milestones according to chronological age which provided the norm for detecting variations or deviations from the expected sequence of behavior.

McGraw contended instead (in the chapter "Behavior Development" included here from her 1935 study) that growth contributes to order as well as the instability characteristic of development. McGraw saw development as a continuous process of interaction between neuroanatomical structures and physiological functions, involving the construction, dissolution, and reintegration of behavior into more varied and complex forms. Development involves functional differentiation and specialization. Growth determines the rate that energy is transformed to affect size and complexity (i.e., degree of functional specificity) from birth to old age. Julian Huxley (1931:216–229) a biologist, demonstrated that early growth is heterochronic in some species; faster growing segments influence the timing of emergence and quality of subsequent features, thus altering the proportionality of anatomic parts and organs. Differences in energy input and the rate or velocity of growth introduced elements of indeterminacy in development, that McGraw believed, made possible the modification of the sequence through which neuromuscular functions are combined to form a pattern of behavior (McGraw 1935:16).

McGraw hypothesized that the degree of integration of human behavior depended upon the relationship between ontogenetic patterns of a recent origin and older phyletic traits. Some of these latter traits exhibit a mature (i.e., fixed pattern) form at birth, soon disappear and thus, exert little influence in the emergence of subsequent behavior. However, other behaviors persist in residual form to be eventually incorporated into a stable pattern. Consequently, McGraw reasoned that if older traits do not necessarily cause the appearance of younger ones (a view consistent with Dewey's contention that relation between antecedent and consequent does not imply a causal relationship) it might be possible to change the sequence in which older and younger traits are combined by altering the rate or velocity of early growth and learning processes. That is why McGraw insisted that observational studies be commenced at birth so that the earliest and most subtle neuromuscular events that account for the development of complex behavior could be detected and manipulated. McGraw attempted to graft together movements associated with different behavioral patterns, just as embryologists transplant cells and tissue, to see what form they would assume within the parameters of a new situation or different environment.

McGraw demonstrated the fruitfulness of challenging the idea that behavior develops in a fixed sequence by showing that phyletic traits previously considered useless proved instrumental in hastening the emergence of their ontogenetic counterparts. For example, the seemingly disorganized swaying of the torso and the alternating movement of legs and arms appearing in the early stages of prone locomotion enabled an infant to swim before being able to crawl (McGraw 1939). Johnny's ability to skate before he could walk smoothly also demonstrated that the rhythmic stepping movements involved in skating could be stimulated to hasten the attainment of erect locomotion (McGraw 1935:160–164).

The importance of Coghill's contribution to Gesell and McGraw's studies cannot be overstated. Coghill (1930a:638) claimed to have discovered a fundamental law governing the development of unconditioned (spontaneous or involuntary movements) and conditioned (e.g., stimulated) reflexes, by contending that "partial [segmental] patterns arise by individuation through the restriction of both the field of motor action and the field of adequate stimulation." For example, a fetal salamander's earliest movements are instigated by endogenous feelings, producing a series of alternating contractions of the trunk that proceed tailward, eventuating in locomotion. When appendages first appear, sensory systems are relatively undeveloped. Consequently, the application of a local stimulus to the limb bud, for example, results in the inhibition of a total pattern, characterized by the suspension of movement and assumption of a fixed postural stance or orientation. Coghill believed that pre-sensory postural mechanisms evoke "primary attitudes" associated with complete neuromuscular integration. As structural counterparts to experience, these attitudes anticipate the acquisition of function by providing the orientation or balance needed to guide initial development. Coghill also believed that these "primary attitudes" continue to exert an important influence in subsequent learning processes (Coghill 1930a:641–642). McGraw devoted considerably more attention to the role of attitudes in developmental behavior than Gesell, and concluded that attitudes constituted the most variable aspect of behavior.

Coghill suggested how McGraw could tap into the mechanisms and attitudes contributing to the coordination and functional integration of behavior, before their suppression with the onset of cortical control. McGraw interpreted Coghill's conception of integrated response experimentally (see Chapter 9) to mean that the problem solving situations she devised must pose challenges that could not be overcome by an infant through some reflexive response, but only by mounting a coordinated response. Thus, when confronted with uncertainty, McGraw thought infants were likely to suppress an immediate response until they got a feeling for the situation as a whole. McGraw challenged infants to extend their proficiency beyond those conditions already under their control by placing them in situations requiring problem solving through inventive motor responses, such as that involved in scaling inclines, climbing off stools, or roller skating. These techniques of special stimulation revealed how infants learn by adopting certain postures and attitudes and drawing out and elaborating the physical movements involved.

When placed in a sitting posture on a stool, for example, most children go through a rather laborious process of rotating to a prone position, inching their way backward, grasping the edge and dropping off. When Johnny was first introduced to this activity, at an earlier age than other infants, he was instructed to take hold of the edge of the stool and turn to one side before throwing himself forward. This technique enabled him to accomplish this task sooner than the others (McGraw 1935:156). However, the task was then made more difficult by having him commence the activity in a standing position. Confronted with the new problem of getting into a balanced sitting position before descending from the stool, Johnny was forced to deal with a more complex situation (i.e., involving a wider field of stimuli), engaging a broader range of potential motor responses. However, in the course of resolving this more challenging problem of balance, Johnny effectively eliminated three subsequent phases in the sequence of dismounting a stool no longer essential to successful performance which included sitting, rotating to a prone position, and edging forward. Johnny eliminated these intermediate steps by using his hands and legs more effectively in one integrated movement involving kneeling, grasping the sides, pivoting, and thrusting the body over the side by pushing with hands and legs (McGraw 1935:157–158). Thus Johnny successfully transferred a technique to overcome a new, more challenging situation.

Johnny's early opportunities to attain a sense of equilibrium or balance in climbing off stools, roller skating, and many other challenging exercises seemed to give him an added poise and self-confidence. He tended to pause and gauge the demands of a problematic situation unlike Jimmy, and devised a more effective sequence of actions to overcome unanticipated difficulties. Johnny also demonstrated more persistence and even temperance. In contrast, Jimmy was more direct but less deliberate. He succumbed to frustration when more complex tasks proved elusive and returned to more rudimentary and laborious methods. Johnny's "acquiescent" attitude, as McGraw characterized it, best exemplified Dewey's contention that the postponement of an immediate response prolongs inquisitiveness, decreases resistance to doubt and ambiguity, and increases openness to the suggestiveness of the situation. Consequently, Johnny was better able to comprehend the requirements of the situation in its entirety, thus increasing his ability to direct and control his behavior (McGraw 1935:285). Moreover, Fentress and McLeod (1988:72) suggest that uncertainty involved in incomplete, ambiguous, or even distorted experiences is more likely to redirect the course of development by forcing individuals to "fill-in" the details, by making conscious choices that clarify the situation as a whole.

Attitudes infants express through gestures in problem-solving activities provided McGraw important clues as to the level of awareness, deliberateness, and degree of learning involved in their completion. Frowning , crying, and other gestures reflected feelings governed by subcortical processes. McGraw (1943a:108–109) contended that an infant's capacity for emotional expression did not emerge until around the last quarter of the first year. During this critical phase, infants exhibited increased motor control that McGraw believed to be indicative of expanded cortical control, such as adjusting to an inverted position, rolling from a prone to supine position, achieving a sitting position, and so forth.

In this regard, Hadders-Algra and Prechtl (1992:210) suggest that an attitude of cheerfulness, reported to accompany infant swiping and swatting at 3 months, may be indicative of the formation of thalmo-cortical connections that enable temperament to be expressed in specific forms of behavior. Moreover, Jerome Kagan and his associates contend that a temperamental predisposition to be inhibited and wary or uninhibited and curious in new situations may be accounted for in part, by neurochemical processes in early development (see Robinson et al. 1992). Interestingly, Gesell promoted the idea (nearly opposite of McGraw's) that children should not be rushed prematurely to attempt activities beyond their motor abilities, for to do so risks creating emotional disorders that will interfere with learning in later development. However, McGraw found no one-to-one correspondence between attitude and capability. For example, overeagerness sometimes interfered with problem-solving, increasing frustration, while pensiveness and restiveness prolonged the search for an appropriate solution.

EXPLAINING DEVELOPMENT

Perhaps McGraw's most important but least understood contribution to knowledge about infant growth was her conception of the patterned matrix of neuromuscular development. Noteworthy is the fact that the *weaving* metaphor she employs in the last chapter of *Growth* (included here) was published four years *before* Gesell advanced his own theory, proposing the principle of "spiral organization of reciprocal interweaving" to explain infant development (Gesell 1939). McGraw (1935:306–308) observed that growth processes are not straightforward. Developing patterns overlap in an interdependent and alternating fashion, pulsating or oscillating forward and backward in rhythmic waves. Development occurs in distinct phases, according to McGraw, involving exaggerated and inhibited movements, the elimination of excess motion and the consolidation and integration of behavior. McGraw and Gesell offer extraordinarily powerful and elegant explanations of the complex dynamics of neuromuscular development that, in many respects, complement one another. The metaphors Gesell and McGraw employ to explain development are quite suggestive, but the phenomena they encompass extend well beyond current concepts and experimental methods available to make their complete elaboration possible. This objective must await further advances in knowledge about the relationship between neural growth and behavior. However, there are several crucial differences in Gesell and McGraw's theoretical perspectives, the contours of which may be depicted graphically, that are not only of historical interest, but involve issues that have a bearing on current research by developmental biologists and psychologists.

There are at least five dimensions in which their respective theories can be usefully compared without introducing overwhelming detail. These dimensions include: (1) the basic unit of analysis; (2) the directional tendency of growth and developmental processes; (3) the neuromuscular phases involved in the construction of behavior; (4) the mechanisms that account for the integration of behavior and learning and; (5) the developmental trend beyond infancy.

The Unit of Developmental Analysis

Gesell argued that an integrated or mature pattern of behavior consists of the "progressive incorporation" of single traits into a more complex pattern, as denoted in Figure 1, by reincorporating a, b, c, and d through successive stages to form d1. Gesell (1945:176) contended that each trait "nascently" embodies the unitary action system from which it was derived but that this state of prior integration cannot be attained until after traits a-d have interacted through successive recombination to form a complete corpus of behavior. Since we are unable to know the exact phyletic sequel in which traits emerged to form a pattern such as prehension, Gesell (1945:280) argued that the choice of a specific trait is arbitrary. (Gesell happens to choose radial raking which appears much later in ontogeny than several other traits that precede it, thus narrowing the full range of traits that make prehension possible.) This is a significant departure from Coghill's conception of prior inte-

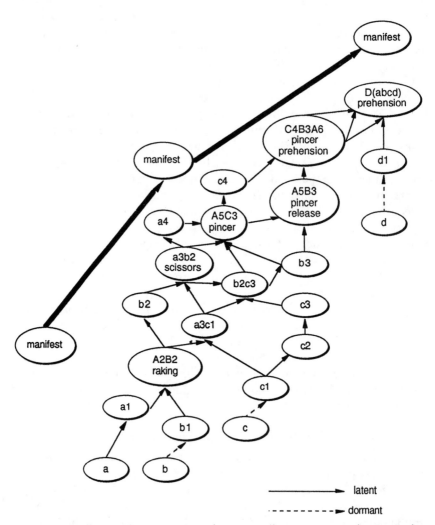

Figure 1. Gesell's Spiral Organization of Reciprocally Interwoven Behavior: Prehension (adapted from A. Gesell, 1945, *The Embryology of Behavior* [New York: Harper Brothers, 174]).

gration, as apparent in subsequent elements of Gesell's developmental theory.

McGraw concluded that the genesis of erect locomotion was more likely than prehension to provide an accurate picture of the sequence of development. Erect locomotion incorporated functional vestiges of older neural structures, such as the righting response that appeared first in ontogeny, and this was consistent with Coghill's demonstration that integration preceded individuation. The transition from prone to erect locomotion involved more than just a recombination of separate movements into a more complex pattern, McGraw believed, but a series of reintegrations involving the displacement of mass and realignment of neuroanatomical structures and behavioral functions into a new equilibrium. In this regard, according to Provine (1988:42–43), changes in muscle mass affect the number of motoneurons competing for sites of innervation. Accordingly, anatomical remodeling, involving increases in muscle mass associated with erect locomotion, could have engaged a whole new set of motoneurons (through interneuron target processes), making bipedal movements possible.

Dewey and McGraw believed that the nervous system converted energy into various forms of motion, exhibiting common attributes of velocity, amplitude, frequency, and direction that stimulated and guided processes of growth and development (see Wetzel 1937 for a more detailed elaboration). Consequently, they thought it theoretically possible to render seemingly incommensurate neural and behavioral events into equivalent quantitative terms. Recent studies of neural growth processes suggest that position, velocity, geometry, and patterned movements of neurons and neuroglial substrate are fundamental to the morphogenesis of the nervous system. For example, locomotion plays an essential role in axon guidance and steering mechanisms (Bray 1991). Axons grow through a complex series of crawling and gliding movements involving adhesion, elongation, and retraction that display characteristic morphologies at different stages of development (Bridgman 1991). These movements may enable axons to respond to external cues and follow pathways that result in adhesion and functional specification. Research on axon growth may help pinpoint the transformational properties of motion essential to behavioral changes that contribute to learning and the capacity for adaptation occurring throughout development (See Jacobson 1992 and Purves 1985).

According to Oppenheim (1982), Coghill was the first to argue that inhibition of peripheral movement plays a fundamental role in fetal motility and subsequent behavioral development. Subsequent experimental research indicates that inhibition may precede excitation in the onset of embryonic motility, evoking spontaneous, rhythmic motions (through successive cycles of flexion through tension and release) that contribute to the subsequent development of neuromuscular systems (Oppenheim 1982:290). The possibility that inhibitory neurotransmitters serve a transient function by regulating cell proliferation, migration, and differentiation has also been given serious attention. Moreover, additional research suggests, according to Oppenheim (1982:291), that inhibitory mechanisms may enable the incorporation of functional adaptations in the nervous system during pre-natal development. Thus, Oppenheim (1988:22) contends that whether inhibition or some other mechanism proves to be the ultimate source of these ontogenetic adaptations, "transient neural adaptations may mediate *both* immediate *and* future needs" of the organism.

As noted before, McGraw believed that some of the earliest but rapidly waning forms of neonatal behavior embodied integrated responses. These behavioral forms governed by the cerebellum served a functionally useful purpose until taken over by the cortex. This assumes, of course, that transient neuromuscular responses truly represent actual ancestral forms, since these responses could have undergone extensive modification in ontogeny (see Jacobson 1992:64). Nevertheless, cell death and regrowth constitute normal features of early neural development, facilitating the extensive modification and reorganization of connections necessary for complex adaptive behavior (Oppenheim 1981; Prechtl 1982). Tilney and Kubie (1931:300–301) contended that cortex evolved in such a way to sustain the influence of these older neural structures in the early stages of development, before sensor-motor systems have reached maturity. This phenomena was evidenced in the layering and lamination processes by which the older archicortex appeared at a later embryonic stage than the neocortex, enabling the afferent projection system to establish a functional connection with a "general body sense" (Tilney and Kubie 1931:309). Subsequent research in cell lineage has disproven the view that the cerebral cortex develops sequentially from the older surface layers inward to the younger core (Jacobson 1992:416). Nevertheless, there is a striking overall alternation in the direction of cerebral growth (as McGraw discovered in behavioral development) between laminar regions that enables the interpenetration of younger and older migrating neurons (Jacobson 1992: 419).

Developmental Direction

Gesell and McGraw also advanced different explanations to account for changes in the direction of behavior growth during development. Gesell contended that the progressive tendency of behavior development is from the *inside-out*; that is, specific traits initially reside in the substratum and manifest themselves only after passing through successive stages of latency. Gesell held that older traits are never directly expressed in manifest behavior but subsumed and reformed through recombinant processes. McGraw did not accept this principle of latency. She believed that all traits must be expressed in behavior, no matter how briefly, for extensive remodeling to occur and eventual neuromuscular integration.

The direction of development is not straightforward, according to Gesell, because combined traits are temporarily split off and diverge from one another (illustrated in Figure 1) as they come within the field of influence of other traits, before recombining to form a more complex pattern. Traits that become separated through divergence look like they are "retreating", according to Gesell (1939:179), but are actually "more mature" because they do not continue on the same trajectory, but spiral back towards the point of departure within the "same region," but at a "higher level." For example, in Figure 1, the pincer or plucking movement depicted by a3b2 represents a higher form of prehension than the more oblique scissors closure represented by a2 b2. Curiously, however, it is difficult to explain how plucking, which Gesell considers a "higher order of performance" than the scissors method is a comparatively less apt or "deft" response (see Gesell 1934a:320). Although Gesell (1934a:279; 320) says that this awkwardness eventually disappears, he is unable to specify how this happens. McGraw believed that the increased functional specificity of behavior came about not simply by adding more complexity, but also as a result of the elimination of inessential movements.

McGraw proposed a somewhat different scheme to explain early development. She contended that neurobehavioral development was characterized by a rhythmic expansion and contraction that was *bidirectional* in three senses. First, neural growth extends initially from the center to the periphery of the brain and spinal cord, allowing younger neurons to migrate through older cortical layers (as suggested by Coghill and Tilney), anticipating the acquisition of function. The exercise of function then reverses the direction of neural growth, stimulating the the growth of afferent fibers back into neural centers governing sensori-motor integration. Second, behavior is initially manifested in an integrated but abbreviated form, dominated by the trunk. Then it is individuated into various segmental forms of activity associated with primary forms of locomotion. The many occasions for behavioral adaptations encountered in the course of development are triggered by disequilibrium (such as the imbalance infants experience in trying to walk) and the felt need to become reoriented to a situation by attaining a stable posture. Consequently, infants and toddlers select that combination of individual movements and postures that best sustains the equilibrium needed to attain new forms of focused and integrated behavior.

Finally, the bidirectionality of development is expressed by the interaction or interdependence of neuroanatomical structures and behavioral functions. McGraw (1940a) humorously illustrates this phenomena most effectively in the attainment of bladder control. Children on the verge of mastering bladder control are frequently unable to express the need to go the the bathroom until after it is too late, because they are only in the early stages of using language, and thus are unable to integrate their verbalization of this need at an appropriate time before voiding. McGraw (1940a:586–587) says that this temporary "diminution of proficiency" occurs because the child's recently established association between the act of voiding and the result is interrupted by having to establish a new association between the feelings or twinges that accompany the urge to void and the need to find the proper receptacle for doing so. In this instance, the feeling of urgency and the capacity to void at will, made possible by the extension of cortical control in the pelvic area, must be controlled and redirected towards the anticipation of finding a bathroom in a timely manner.

Neuromuscular Phases of Behavior

Gesell believed that (1939:169) two primary features of neuromuscular processes give developmental behavior a distinctive form: the intrinsic opposition between flexor and extensor movements and the demand for symmetry between them. Charles Sherrington (1961) first discovered that reciprocal innervation is a precondition for coordinated movement of antagonistic muscles; one set must be inhibited while the other is employed. It should be noted that the principle of reciprocal innervation primarily applies to the movements of involuntary reflexes under sub-cortical control. Nevertheless, Gesell advanced this principle to account for the shifting ascendencies of symmetries involved in the acquisition of a preferential aptitude, illustrated by right or left-handedness. Environmental influences also played a significant role by "inflecting" or bending a developmental trend towards a specific unilateral preference. Although Gesell failed to clearly illustrate how inflection works, he contended, like McGraw, that postural attitude exerts a significant influence or bias in shaping aptitude. Gesell (1939:180; 1945:193) generally held that structured neuromuscular tensions determined the overall trend or directionality and the hierarchical continuum of development. But he left unanswered the crucial issue of what role if any, the cortex plays in early developmental processes.

McGraw attempted a more ambitious inquiry than Gesell by tracing the complex role of the brain in developmental behavior. She believed that cortical inhibition was not only essential to voluntary action but was indispensable to the formation of neuronal connections necessary for the coordination of feeling, movement, thought, and action. McGraw (1940b:1038) found evidence that cortical inhibition spreads *selectively* by alternating between upper and lower regions of the body, gradually altering the extent and range of movement in each region or segment. The lack of a consistent direction and form in developmental behavior was due, in part, to differences in velocity of the growth of neuroanatomical structures. For example, McGraw suggested that early forceful but ephemeral, rhythmic bursts of neonatal movements serve primarily to direct neural growth, facilitating changes in the scope and duration of neural connections needed to support optimum development of a particular motor pattern.

Importantly, McGraw found that the period of time before *offset,* or completion of a neuromuscular phase, particularly later ones involving more complex coordinations (as illustrated in Figure 2 by phases 4–9) was considerably longer than for those occurring at the earlier stages. Infants also showed less variation in the onset and form of rudimentary behaviors involved in creeping for example, such as rhythmic flexion and extension, than in modes of upper and lower body propulsion. Marked variation also occurred in the assumption of preliminary quadrupedal postures (phase 7) for deliberate but unorganized progression in the transition from creeping to crawling (McGraw 1941:101; 105). McGraw's perspective on locomotion contrasted sharply with Gesell (1939:168) who contended that prone locomotion consisted of four distinct postural phases that included frogging, unilateral knee thrust, pivoting, and planigrade progression. According to McGraw, some infants successfully propelled themselves on their stomachs before attaining a creeping posture, for example (see Figure 2, phases 4–5), by pulling with arms and pushing with legs (a method of propulsion that combined elements of frogging and knee thrust), while others were unable to move until after attaining a creeping posture. Thus the form of a behavior pattern was partially determined by the rate of advance of cortical control, the strength and duration of behavioral precursors, and the degree of individual awareness in exploiting the available repertoire of movements.

In this regard, Steven Roberton (1988:89) contends that cyclic spontaneous movements of neonates may help to "segregate beneficial but incompatible processes" (i.e., involving different biochemical conditions) and that ultimate knowledge of the neural mechanism may help isolate how attention functions to connect thought with action (Roberton 1993:685). Prechtl and his associates are pursuing a similar line of inquiry in the analysis of pre-natal and post-natal behavior. Prechtl (1989) determined that the overlapping development of pre-term infant behaviors may facilitate the "synchronization" of movements observed in the transition in behavioral stages of development. Hadders-Algra et al. (1992) also found that general spontaneous movements (i.e., complex and variable motor patterns involving limbs, head, and trunk) involve the co-activation (not reciprocal innervation) of antagonistic muscle groups and that this co-activation persists from "writhing" to "fidgety" patterns during the first 8 weeks of infancy. They argue that the cyclic alternation between upper arms and upper legs and lower arms and lower legs involved in writhing, fidgeting, and later movements (i.e., "swipes" and "swats") indicates that motoneurons calibrate the proprioceptive system by providing feedback essential to motor coordination (Hadders-Algra et al. 1992:248). Their research also leaves open the possi-

142

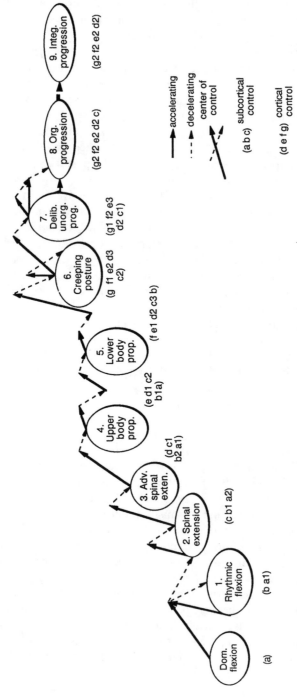

Figure 2. McGraw's Growth Curve of Patterned Behavior: Prone Progression (adapted from M.B. McGraw, 1941, "Development of Neuromuscular Mechanisms as Reflected in the Crawling and Creeping Behavior of the Human Infant," *Journal of Genetic Psychology* 58:83–111).

bility that selective cortical inhibition may help account for this alternating
pattern of general movements, as first described by McGraw.

As the velocity of neural growth stimulating a particular functional complex subsides, the development of other behavioral segments accelerate, creating new centers of dominance and the exaggerated or excessive exercise associated with the acquisition of a new capability. The cross-pressures
entailed with varying growth rates of functional attributes frequently result
in the regression towards more rudimentary forms of behavior. McGraw argued that the point of intersection of accelerating and decelerating behaviors
constituted a critical juncture in integration of behavior, involving two related events illustrated in Figure 2. Each new trait in a behavior complex
(e.g., 1,2,3 etc.) contributing to prone locomotion for example, assumes the
leading position in the sequence, subsequently displacing prior traits to the
end of the series, where their influence is curtailed (i.e., traits d-g) and/or
eventually eliminated (traits a-c). Consequently, excess activity governed by
the subcortex (traits a-c) is inhibited with the onset of cortical control in the
transition from phase 3 to 4 and restricted to that amount necessary to attain
a temporary equilibrium, before the locus of behavior shifts to new centers
of (cortical) control. This checks the regression or further diminution of previously attained capabilities until a new integration is attained by recombination through substitution and displacement. Unlike Thelen and Ulrich
(1991), McGraw did not believe that such behavioral transitions could be
explained solely in terms of a shifting center of gravity associated with physical growth, but also by the degree of awareness and deliberation infants exhibited in adopting postures that restored equilibrium, facilitating the recombination of opposing movements in some new integrated form.

McGraw carefully stressed that complex patterns associated with voluntary or deliberate behavior always retained some degree of subcortical influence through behavioral vestiges that provided the slack for subsequent
modifications in ontogeny. Provine (1988:43) even speculates that the neural circuitry underpinning archaic motor patterns may simply be suppressed
and could be reactivated to "increase the library of potential behavior from
which future motor scores may be composed."

Mechanisms of Integration

Gesell singled out the tonic-neck reflex (t.n.r.) as the single most important
mechanism governing postural set and attitude involved in attainment of
neuromuscular maturation. The t.n.r. provides the earliest indicator of the
capacity for balance and orientation that is an essential precondition to subsequent integration, including the capacity for learning. Gesell (1939:177)
discovered that the t.n.r. was exhibited nearly universally by infants in the
fourth week of prenatal life when observed in a free supine posture. The
head is characteristically rotated to one side toward the extended arm and

leg while the other arm and leg are flexed. Gesell (1938) had occasion to relate his discovery to Coghill and solicit his reactions about its potential significance. Coghill responded in an intriguing way by observing that: "There
is no rotation of the head in Amblystoma so there cannot be typical t.n.r.,
but it seems to me that, while the face is rotated toward the extended arm in
the human infant, the head as a whole is bent away from it. I am wondering
if that could be true" (Coghill 1938:2).

This seemingly innocuous observation is quite significant for it highlights
an important difference between Gesell and McGraw's explanation of a pivotal event in developmental behavior. Gesell (1945:58) contended that the
t.n.r. was a "landmark in the ontogenesis of behavior" because it embodied
a visual-manual-prehensory posture, thus foreshadowing the infants eventual transcendence of the "limitations of the archaic motor system." Coghill
was tactfully alluding to an alternative explanation which maintained the
primacy of the motor system, by suggesting that bending away of the head
from the extended arm was governed by flexion of the trunk and that therefore, a prehensory posture was made possible by partial inhibition of an integrated response. In fact, Coghill (1939) described his last major research
project in correspondence with Gesell that he believed would settle the controversy about the origin of reflexes—a project that led Coghill to argue that
flexion constituted the primary form of response throughout early development (see Coghill 1943:485). (Incidentally, Thelen and Ulrich (1991:89–90)
seem unaware that their experimental analyses actually corroborate
Coghill's thesis by demonstrating the dominance of flexor muscles in infant
stepping and early walking.)

Similarly, McGraw argued that the functional significance of the t.n.r.
posture could only be demonstrated within the context of the overall pattern
of behavior in which it was embodied. McGraw (1943a:43–44) believed
that the turning of the head and face toward the extended arm, preparatory
to rolling from a prone to supine position constituted at one time, integrated
elements of a "righting reflex" subsequently separated during the evolution
of erect posture. Consequently, McGraw concluded that the t.n.r. does not
presage eye-hand coordination but simply foreshadows the preliminary postural adjustments (i.e., coordination of shoulder and hip movements) needed
to master prone locomotion. Only later, does vision become important in
enabling the infant to judge distance correctly to crawl, and ultimately, to attain erect locomotion (McGraw 1943a:47–48).

McGraw had some important reasons, based on her experimental research, for rejecting the conventional wisdom that learning begins when an
infant recognizes and is able to grasp and manipulate objects. Her studies of
reaching and prehension indicated that an infant's early fixation on and pursuit of objects did not necessarily entail visual recognition and identification
(McGraw 1943a:93–101). Fixation may reflect poor coordination of eye

movement and reaching does not necessarily imply an attempt to grasp. McGraw argued instead, that when closely examined, these behaviors enable an infant to learn to accommodate distance by adjusting movements for convergence on objects. For example, infants gauge distances initially by reaching close fisted toward an object, with the mode of grasping (i.e., scissors or pincer) eventually adopted being determined primarily by the size and shape of the object. The appearance of "pointing" demarcates increased awareness, according to McGraw, because it reflects the infant's ability to inhibit compulsive reaching and deliberate before seizing an object (see Touwen's afterword for additional observations on the duration and significance of pointing).

Recent evidence from neurobiology however, supports both McGraw and Gesell. Visual experience is not required for feature detection because the neonatal cortex may be already organized to respond selectively to orientation at birth (Purves 1985:354). Although experience does not literally instruct the neuromuscular system how to attain a final pattern, it does contribute to the reorganization of synaptic connections during early development that affect the form of behavior (Purves 1985:275–280; Oppenheim 1981). However, post-natal vision may play a decisive, instructive role at some point between shaping peripheral and associational connections (Purves 1985:354). Nevertheless, the function of learning, whether generated by events within growing bodies or by external stimuli is roughly the same, according to Purves (1988:168), to insure neural capabilities will be adequate to meet uncertain and unforeseen circumstances.

Long Term Developmental Trends

Gesell's longitudinal studies were limited to the first two years so he was not able to address the long-term implications of his developmental theory in any definitive way. Nor was he able to explore in depth the complex and interrelated neural and physiological processes that contribute to a deeper understanding of the relationship between biological growth and development. Gesell (1945:10) hypothesized that developmental trends were governed by three principal mechanisms which included respiratory rhythms, tonicity, and "electrotonic integration" (or the gradient or field relations between neurons governing innervation). As noted in the general introduction, generous multiyear foundation grants and strong support from Frederick Tilney and John Dewey, including an interdisciplinary team of scientists, made it possible for McGraw to explore the larger dimensions of developmental processes during infancy. McGraw summarized the findings of her advanced studies in the chapter of "Individual Development" in her book, *The Neuromuscular Maturation of the Human Infant,* 1943. (For a complete bibliography of advanced studies done by McGraw and her associates in the Normal Development Child Study from 1936–1943 see Weech, et al., 1960).

McGraw and her associates attempted to predict individuality or that point at which each infant had attained neuromuscular integration of a series of different behaviors. McGraw's associates Campbell and Weech (1941) computed each child's development according to three separate measures: extent of neuromuscular maturation at birth; rate of attainment of an integrated motor activity; and age of achievement of a motor skill. Although rate or velocity of change was a better indicator than maturation at birth in predicting individuality within the first one-hundred and fifty days, age of achievement constituted the best predictor of the ultimate rank order occupied by each child (McGraw 1943a:116). Fluctuations in the velocity and direction of development made it impossible to select a comparable time period for two or more individuals unless individual attainment was computed as an average of scores for several activities. Therefore, McGraw suggested that inter-individual comparisons could be drawn if each child's developmental history was laid side by side like ladders so that behavioral milestones could be compared according to rate as well as trajectory. This method, according to McGraw, yielded more accurate predictions of individuality than rank order of behavior milestones by age alone.

McGraw and her associates also mounted a series of investigations to see if an acceleration constant could be devised that would render neural and behavioral phenomenon comparable that are moving at different velocities and in contrasting directions during development. These and other exploratory investigations were of utmost importance to Dewey (as noted in the general introduction) who sought to demonstrate in *Logic* that judgment is a psychobiological capacity to maintain a balanced perspective, enabling the investigator to convert incommensurate or discontinuous phenomena into common forms, thus allowing the processes of inquiry to go forward.

For example, McGraw and Breeze (1941:291–295; 299–301) concluded that the development of erect locomotion involves a redistribution of energy to support the anatomical and postural adjustments needed to attain bilateral symmetry and that the energy required for a stable gait in walking decreases with age. A. C. Weinbach (1938), a biochemist, found similar evidence that an infant's proficiency in climbing slides increased with age and that power output is adjusted (i.e., decreased) to reflect the child's increased efficiency and competence. In addition, psychologist Vera Dammann (1941) demonstrated that infants' increased proficiency climbing slides also led to the adoption of positive attitudes, indicative of greater self-confidence. These studies suggested that learning involves progressive advances in human behavior, supporting Dewey's contention that judgment transforms human experience by converting energy into a form that facilitates change. (For a concise presentation of principles of biomechanics, see Winter 1990.)

Finally, Roy Smith (1938), a neurophysiologist, argued that the presence of synchronous electrical brain activity in neonates demonstrated the waning influence of a phylogenetically older pacemaking mechanism before the onset of cortical control. This finding lent added credibility to Tilney's thesis that development involves the reciprocal interaction of older and younger regions of the brain. His studies also suggested that cerebral rhythms provide the balance infants need in early development to exercise judgment despite undergoing continuously destabilizing growth and change.

Weinbach (1938) formulated an equation showing how the rate of frequency of brain waves measured by Smith was correlated with a similar proportionate increase in neural cell mass and weight (that declined with age), suggesting the possibility that mechanisms controlling the velocity of growth are present at birth. Weinbach (1941a; 1941b) devised a series of equations to calculate the effects of physiological growth involving reciprocal adjustments between the velocity of growth (i.e., rate of energy transformation) of the brain and behavior. He derived an acceleration constant from these equations that predicted the trajectory of human development from fetus to old age. Significantly, the "S" shaped human growth curve across the life span exhibited the characteristic alternating sequence of accelerating and decelerating phases first observed by McGraw.

For example, if the human growth curve in Figure 2 represented the lifespan, it would include a period of pre-natal growth (a-1); a period extending from birth to puberty involving brief periods of acceleration and deceleration (2-5); followed by a period of increased acceleration through adolescence (6-7); and, finally, a decelerating phase continuing through old age (8-9). This pathbreaking work by McGraw's associates convinced Dewey and McGraw that behavioral equivalents could be found for biological phenomena, making possible the conversion of complex developmental processes into symbolic form (see Dalton and Bergenn 1994).

SOCIAL INTEGRATION

The last article included in this part, *Let Babies Be Our Teachers,* presented by McGraw as a Harvey Lecture for the New York Academy of Medicine in 1943, underscores the profound impact of Coghill's ideas on McGraw's thinking about infant development, and the striking parallels they suggested between biological organization and processes of social development. This article served as a point of departure in McGraw's subsequent effort throughout the remainder of her life to articulate the social and cultural ramifications of her research in child development. Although a complete elaboration of her social theory is beyond the scope of this introduction, (see Dalton, Bergenn, and Lipsitt 1995) three themes foreshadowing her later work deserve comment. First, McGraw (1943b) believed that social systems grow and evolve and are confronted with the same problems of organization, communication, and learning that confront the infant. The world at war in the 1940s dramatized the fragility of world stability and the overriding need for integrated solutions to disputes about national sovereignty. McGraw thought Coghill's principle that integration preceded individuation was eminently suited to govern the world in the post-war era where national sovereignty was acknowledged and defended in exchange for international cooperation and the renunciation of the use of force.

Second, McGraw believed that societies like infants could learn from experience and evolve behaviors and adopt roles that contribute to the progressive enlargement of human experience. Social systems faced essentially the same challenges that confronted neuromuscular systems: the need to achieve cohesion through the effective coordination and harmonization of individual capacities. Social cohesion is no more instantaneous than is the attainment of walking; both require a conscious and deliberate attempt to overcome imbalance, insecurity, and uncertainty to step forward and forge integrated and unified relations based on trust, shared goals, and mutual respect. McGraw believed humans could acquire a "social cortex" or social consciousness capable of overcoming divisive, regressive forces favoring fear and impulsiveness over deliberation and planning.

Finally, McGraw believed that communication was as instrumental to social development as it was to individual learning because it served as the medium through which attitudes are expressed and embodied in behavior. The learning process is easily stymied, as McGraw discovered, when negative attitudes overwhelm the ability to respond to new challenges. Similarly, the possibility of social change and integration is thwarted to the extent that the norms governing individual roles embody attitudes that favor conformity, create suspicion, and promote intolerance rather than generate feelings of solidarity and respect for differences.

References

Ames, L.B. 1989. *Arnold Gesell: Themes of His Work.* New York: Human Sciences Press.

Bergenn, V.W., T.C. Dalton, and L.P. Lipsitt. 1992. "Myrtle B. McGraw: A Growth Scientist." *Developmental Psychology* 28:381–389.

Bray, D. 1991. "Cytoskeletal Basis of Nerve Axon Growth." In *The Nerve Growth Cone,* eds., P.C. Letourneau, S.B. Katen, and E.R. Macagno. New York: Raven Press.

Bridgman, P.C. 1991. "Functional Anatomy of the Growth Cone in Relation to Its Role in Locomotion and Neurite Assembly." In *The Nerve Growth Cone,* eds., P.C. Letourneau, S.B. Katen, and E.R. Macagno. New York: Raven Press.

Campbell, R.V.D. and A. Weech. 1941. "Measures which Characterize the Individual During the Development of Behavior in Early Life." *Child Development* 12:217–236.

Chaney, B.and M.B. McGraw. 1932. "Reflexes and Other Motor Activities of the Newborn Infant." *Bulletin of the Neurological Institute of New York* 2:1–56.

Coghill, G.E. 1929. *Anatomy and the Problem of Behavior.* New York: Hafner.

_____. 1930a. "The Structural Basis of the Integration of Behavior," *National Academy of Sciences Proceedings* 16:637–643.

_____. 1930b. "Individuation Versus Integration in the Development of Behavior," *Journal of Genetic Psychology* 3:431–435.

_____. 1931. Letter to C.J. Herrick. (October 14). Neurology Collection, C. Judson Herrick Papers, Spencer Research Library, University of Kansas, Lawrence, Kansas.

_____. 1933a. "The Neuroembryonic Study of Behavior: Principles, Perspectives and Aims." *Science* 78:131–138.

_____. 1933b. Letter to C.J.Herrick. (October 27). Neurology Collection, C. Judson Herrick Papers, Spencer Research Library, University of Kansas, Lawrence, Kansas.

_____. 1934. Letter to A. Gesell. (April 23). General Correspondence, Arnold Gesell Papers, Manuscript Division, Library of Congress, Washington, D.C.

_____. 1936. "Integration and the Motivation of Behavior." *The Journal of Genetic Psychology* 48:3–19.

_____. 1938. Letter to A. Gesell. (May 25). General Correspondence, Arnold Gesell Papers, Manuscript Division, Library of Congress Washington, D.C.

_____. 1939. Letter to A. Gesell. (June 30). General Correspondence, Arnold Gesell Papers, Manuscript Division, Library of Congress Washington, D.C.

_____. 1943. "Flexion Spasms and Mass Reflexes in Relation to the Ontogenetic Development of Behavior." *The Journal of Comparative Neurology* 79:463–486.

Dalton, T.C. and V.W. Bergenn. 1994. "John Dewey, Myrtle McGraw and the Logic: An Unusual Collaboration in the 1930s," forthcoming.

Dalton, T.C., V.W. Bergenn, and L.P. Lipsitt, eds. 1995. *Reflections in the Mirror of Childhood: Essays by Myrtle McGraw on the Evolving American Culture and Society, 1940–1980.* forthcoming.

Dammann, V.T. 1941. "Developmental Changes in Attitudes as one Factor Determining Energy Output in a Motor Performance." *Child Development* 12:241–246

Fentress, J.C. and P.J. McLeod. 1988. "Pattern Construction in Behavior." In *Behavior of the Fetus,* eds., W.P. Smotherman and S.R. Robinson. Caldwell, NJ: The Telford Press.

Gesell, A. 1934a. *Infant Behavior: Its Genesis and Growth.* New York: McGraw-Hill.

_____. 1934b. Letter to G.E.Coghill. (March 22). Arnold Gesell Papers, Manuscript Division, Library of Congress, Washington, D. C.

_____. 1938. Letter to G.E. Coghill. (May 17) Arnold Gesell Papers, Manuscript Division, Library of Congress, Washington, D.C.

_____. 1939. "Reciprocal Interweaving in Neuromotor Development." *The Journal of Comparative Neurology* 10:161–180.

_____. 1945. *The Embryology of Behavior.* New York: Harper Brothers.

Hadders-Algra, M. and H.F.R. Prechtl. 1992. "Developmental Course of General Movements in Early Infancy. I. Descriptive Analysis of Change in Form." *Early Human Development* 28:201–213.

Hadders-Algra, M., L.A. Van Eykern, A.W.J. Klip-Van den Neiuwendijk, and H.F.R. Prechtl. 1992. "Developmental Course of General Movements in Early Infancy. II. EMG Correlates." *Early Human Development*. 28:231–251.

Herrick, C.J. 1949. *George Ellett Coghill: A Naturalist and Philosopher*. Chicago: University of Chicago Press.

Huxley, J. 1931. *Problems of Relative Growth*. New York: Dial Press.

Jacobson, M. 1992. *Developmental Neurobiology*. New York: Plenum.

McGraw, M.B. 1932. "Grasping in Infants and the Proximo-Distal Course of Growth." *Psychological Review* 40:301–302.

_____. 1933. "The Function of Reflexes in the Behavior Development of Infants." *The Journal of Genetic Psychology* 42:209–216.

_____. 1935. *Growth: A Study of Johnny and Jimmy*. New York: Appleton-Century. (reprint, New York, Arno Press, 1975).

_____. 1939. "Swimming Behavior of the Human Infant." *Journal of Pediatrics* 15:485–490.

_____. 1940a. "Neural Maturation as Exemplified in the Achievement of Bladder Control." *Journal of Pediatrics* 16:580–590.

_____. 1940b. "Neuromuscular Mechanisms of the Infant: Development Reflected by Postural Adjustments to an Inverted Position." *American Journal of Diseases of Children* 60:1031–1042.

_____. 1941. "Development of Neuromuscular Mechanisms as Reflected in the Crawling and Creeping Behavior of the Human Infant." *Journal of Genetic Psychology* 58:83–111.

_____. 1943a *The Neuromuscular Maturation of the Human Infant*. New York: Columbia University Press.

_____. 1943b. "Let Babies Be Our Teachers." *March of Medicine* ed. New York Academy of Medicine. New York: Columbia University Press.

_____. 1979. Letter to R. Oppenheim. (December 31). Myrtle B. McGraw Papers, Millbank Memorial Library, Teachers College, Columbia University, New York.

_____. 1980 "Growing Up With and Without Psychology." Unpublished Manuscript: 1–65, Myrtle B. McGraw Papers, Leonia, New Jersey.

McGraw, M.B. and K. Breeze. 1941. "Quantitative Studies in the Development of Erect Locomotion." *Child Development* 12:267–303.

Oppenheim, R.W. 1978. G.E. Coghill (1872-1941): Pioneer Neuroembryologist and Developmental Psychobiologist." *Perspectives in Biology and Medicine* 22:45–64.

_____. 1981. "Ontogenetic Adaptations and Retrogressive Processes in the Development of the Nervous System and Behavior: A Neuroembryological Perspective." In *Maturation and Development: Biological and Psychological Perspectives*, eds., K.J. Connolly and H.F.R. Prechtl. Philadelphia: Lippincott.

_____. 1982. "The Neuroembryological Study of Behavior: Progress, Problems, and Perspectives." In *Current Topics in Developmental Biology, Vol. 17: Neural Development*, ed. R.K. Hunt. New York: Academic Press.

_____. 1988. "Ontongenetic Adaptations in Neural and Behavioral Development: Toward a More 'Ecological' Developmental Biology." In *Fetal Neurology*, eds., A. Hill and J.J. Volpe. New York: Raven Press.

_____. 1992. "Pathways in the Emergence of Developmental Neuroethology: Antecedents to Current Views of Neurobehavioral Ontogeny." *Journal of Neurobiology* 23:1370–1403.

Prechtl, H.F.R. 1982. "Regressions and Transformations During Neurological Development." In *Regressions in Mental Development: Basic Phenomena and Theories,* ed. T. Bever. Hillsdale: Erlbaum.

_____. 1989. "Fetal Behavior." In *Fetal Neurology* eds. A. Hill and J.J. Volpe. New York: Raven Press.

Provine, R.R. 1988. "On the Uniqueness of Embryos and the Difference It Makes." In *Behavior of the Fetus,* eds., W.P. Smotherman and S.R. Robinson. Caldwell, NJ: The Telford Press.

Purves, D. 1985. *Principles of Neural Development.* Sunderland MA: Sinauer.

_____. 1988. *Body and Brain: A Trophic Theory of Neural Connections.* Cambridge: Harvard.

Razel, M. 1985. "A Reanalysis of the Evidence for the Genetic Nature of Early Motor Development." In *Advances in Applied Developmental Psychology,* ed., E. Sisel. Norwood, N.J.: Ablex. Vol. 1:171–211

_____. 1988. "Call for a Follow-up Study of Experiments on Long-Term Deprivation of Human Infants." *Journal of Perceptual and Motor Skills* 67:147–158.

Roberton, S.S. 1988. "Mechanism and Function of Cyclicity in Spontaneous Movement." In *Behavior of the Fetus,* eds., W.P. Smotherman and S.R. Robinson. Caldwell, NJ: The Telford Press.

_____. 1993. "Probing the Mechanism of Oscillations in Newborn Motor Activity." *Developmental Psychology* 29:677–785.

Robinson, J.L., J.S. Reznick, J. Kagan, and R. Corley. 1992. "The Heritability of Inhibited and Uninhibited Behavior: A Twin Study." *Developmental Psychology* 28(6):1030–1037.

Sherrington, C. 1961. *The Integrative Action of the Nervous System.* New Haven: Yale University Press.

Smith, J. Roy. 1938. "The Electroencephalogram During Normal Infancy and Childhood: The Nature of Growth of Alpha Waves." *Journal of Genetic Psychology* 53:455–469.

_____. 1939. "The Occipital and Pre-Central Alpha Rhythms During the First Two Years." *Journal of Psychology* 7:223–227.

Thelen, E. and K.E. Adolf. 1992. "Arnold L. Gesell: The Paradox of Nature and Nurture." *Developmental Psychology* 28:368–380.

Thelen, E. and B. Ulrich. 1991. *Hidden Skills.* Chicago: University of Chicago Press.

Tilney, F. and L. Kubie. 1931. "Behavior in its Relationship to the Development of the Brain." *Bulletin of the Neurological Institute of New York* 1:226–313.

Weech, A.A., H. Alexander, D.S. Damrosch, R.L. Day, C.M. Riley, W.A. Silverman, eds. 1960. *The McIntosh Era at Babies Hospital, 1931-1960.* New York: Babies Hospital.

Weinbach, A. P. 1938a. "Some Physiological Phenomena Fitted to Growth Equations III. Rate of Growth of Brain Potentials (Alpha Frequency) Compared with Rate of Growth of Brain." *Growth* 2:247–251.

_____. 1938b."Some Physiological Phenomena Fitted to Growth Equations IV: Time and Power Relations for a Human Infant Climbing Inclines of Various Slopes." *Growth*4:123–134.

_____. 1941a. "The Human Growth Curve: I. Prenatal." *Growth* 5:217–233.

_____. 1941b. "The Human Growth Curve: II. Birth to Puberty." *Growth* 5:233–255.

Wetzel, N.C. 1937. "On the Motion of Growth. XVII: Theoretical Foundations." *Growth* 1:6–59.

Winter, D.A. 1990. *Biomechanics and Motor Control of Human Movement.* 2nd ed. New York: Wiley.

❧ 8 ☙

The Function of Reflexes in the Behavior Development of Infants

The ultimate objective of all psychology is to further an understanding of human behavior, complex human behavior, as it is known in present-day society. Early psychology followed the lead of more established science in attempting to determine the simplest elements and to construct concepts of behavior as it aggregated and became more complex. The simplest types of behavior were called reflexes and elementary textbooks in psychology customarily contained a list of these simple reflexes, another list of slightly more complicated responses, called "instincts" and, upon these inherited, relatively unmodifiable, traits acquired characteristics presumably developed. The identifying characteristics of a reflex as given in some of the more common textbooks are: (a) that they are simple stimulus-response units, (b) that they involve a direct route through the nervous system, (c) they are very prompt in their reaction, specific, and predictable, (d) they have little or no inhibitive or modifiable characteristics and they usually persist throughout life (Gates 1930:627). Woodworth (1921:580) says, "They are quick, definite, given to specific response, involuntary, and often unconscious; permanent and inherent within the organism; they are unlearned and always ready for action." Dashiell (1928:xviii), discussing the same subject, says,

> The elementary action unit into which all behavior can be broken down is that sensorimotor function called a reflex action—or better reflex reaction ... the name hints at the simplicity and the promptness with which this kind of response follows upon stimulation, [however], the majority of reflexes are to some degree compound; several sensory impulses combine to produce the reaction, which may itself be multiple, and thus several different arcs work in cooperation (Dashiell 1928:588.

Reprinted from *The Pedagogical Seminary and Journal of Genetic Psychology,* 42(2):209–216, 1933. Reprinted with permission of the Helen Dwight Reid Educational Foundation. Published by Heldref Publications, 1319 Eighteenth St., N.W., Washington, D.C. 20036-1802. Copyright © 1933. Reprinted by permission of Mitzi Wertheim.

These definitions are taken from three widely used textbooks in introductory courses of psychology. They all stress the simplicity, the specificity, and the promptness of the reflex action, and they all at least imply that an understanding of our complex behavior has its beginning in an understanding of these more elementary reactions. Dashiell (1928:588) makes a point of this when he says,

> What is found true of the simple unit we may expect to find holding true of larger action units as well, and by becoming familiar with the principles as they reveal themselves on a simpler plane we should be able the more readily to recognize their operation in a man's behavior, however complex or however subtle …. the difference is only a difference of degree and in the present chapter we are bearing in mind only the scientific rule of explaining the complex in terms of the simple.

This old reliable scientific principle has, however, in the past few years received considerable challenge and there is a growing tendency, particularly in the biological sciences, to talk in terms of "organismal" reactions rather than elemental. Ritter and Bailey (1927:308), discussing the present assimilation of the idea of unification in all fields of scientific endeavor, say: "In the natural sciences the idea has become established on numerous masses of objective reality highly diverse in character and remote from one another. The most definite outgrowth of the conception as applied to man is found in the extent to which present-day clinical medicine and educational theory are recognizing the importance of the "whole man" and "the whole child." In the field of psychology particularly the Gestalt school has done much to further the idea of "wholeness" especially with respect to the nature of perception, and Koffka (1924:111–112), writing on the nature of reflexes and instincts, says:

> If, by emphasizing the common characteristics of both types of behavior we can now close the gap which previously seemed so wide between the instincts and the reflexes, this does not signify a return to the point of view that instincts are chained reflexes. On the contrary we have reversed the procedure; for it is no longer the reflexive mechanism which is the fundamental fact of behavior, but the characteristic of 'closure' as they appear most clearly in the instinctive activities.

Coghill (1929a) after intensive study of the neuro-functional development of the amblystoma has evolved a thesis of *individuation* or a progressive restriction of zones of adequate stimulation out of a total integrated reaction-pattern. Referring to the elementary characteristic of the reflex, he contends that:

> If there is such a thing as 'unit-reaction' in nervous function it is the total pattern; and the development of specific nervous function, such as reflexes of dif-

ferent grades, is an analytic process not a synthetic one. In so far as the development of behavior is known in vertebrates, all reflexes emerge as partial or local patterns within an expanding or growing total pattern that normally from the beginning is perfectly integrated. They become partial or local only overtly (Coghill 1929:1009).

Lashley (1930), who has approached the problem of behavior through studies of cerebral localization, unequivocally disclaims the usefulness of reflex theory as a key to understanding human conduct. He states:

> In the study of cerebral functions we seem to have reached a point where the reflex theory is no longer profitable either for the formulation of problems or for an understanding of the phenomena of integration. And if it is not serviceable here, it can scarcely be of greater value for an understanding of the phenomena of behavior. ... I believe that there is ample evidence to show that the units of cerebral function are not single reactions, or conditioned reflexes as we have used the term in America. ... The nervous unit of organization in behavior is not the reflex arc, but the mechanism, whatever be its nature, by which a reaction to a ratio of excitement is brought about (1927:12; 17–18).

These quotations represent controversial points of view as to the underlying principles involved in the mechanism and the development of behavior. The bulk of the experimental evidence supporting these contentions has been in the field of animal psychology, though recent studies of infant behavior have essayed to interpret their findings in terms of these general principles. The early studies of infant behavior were limited largely to studies of reflex activities, and, since it was known that the cortex of the newborn infant was in a large measure unmedullated and, presumably, non-functioning, this seemed logical. But the relationship of these reflex reactions to maturation and learning has been given little scientific consideration. In general, most child psychologists and educators have adhered to the chain-reflex theory of development. They have assumed that infants are born with certain specific reflexes and, by a process of conditioning, complex human behavior is fashioned. Watson's early work on the primary behavior equipment of newborn infants and its development by a process of conditioning is familiar to all students of psychology. The bulk of the experimental studies of infants has followed this general principle, viz., that behavior develops from relatively simple reflex-patterns to complex integrated wholes. Some more recent writers have taken the stand that there are very few specific reactions to definite stimuli in the behavior repertoire of the newborn infant. Pratt, Nelson, and Sun (1930) contend that:

> The infant at birth represents an organism in which differentiation has proceeded to the point where there are many effectors and many receptors. Its behavior, however, is generalized. That is, stimulation of almost any group of receptors by almost any kind of stimulus will lead to a response in almost any part

of the organism. The reaction tends, however, to manifest itself most strongly in that part of the organism which is stimulated, and from there spreads out with decreasing frequency and intensity to other segments of the body. This does not mean that the activity within any given segment is well coordinated. ... The newborn infant is equipped with quite a number of reflexes, but the degree of their specificity and their significance seem to have been unduly exaggerated (ix).

Shirley (1931a:227) basing her conclusions on an intensive study of 25 infants over a period of two years, endorses Coghill's theory of individuation. She remarks that, in the human infant at birth,

Individuation of some reflexes has already occurred. Nevertheless, the sudden appearance of integrated locomotor skills that the babies apparently had never practiced is in accordance with the development of locomotion in amblystoma. The law of integration first and individuation into reflexes later probably applies to babies as well as to lower vertebrates. Certainly it is impossible, even by prolonged and careful observation, to see the building up of locomotion from reflexes.

These quotations are taken from two recent publications concerning infant behavior and they indicate a trend of child psychologists to renounce the chain-reflex theory of the development of behavior.

There is in the behavior development of the infant evidence supporting both of these theories. Certainly, the early prancing or walking movement of the newborn infant is a localized segmental reflex functioning at birth. Irrespective of the theories concerning the process of behavior development, it is certainly safe to assume that those behavior patterns which manifest themselves soon after birth are innate in character and they probably have a definite ascertainable relationship to the course of development. It matters little whether the genetic development of human conduct is an aggregation of connections or a process of individuation, the primary problems confronting the infant psychologist are: (1) the determination of both specific and general reactions of newborn infants, and (2) the interpretation of the relationship between these early infantile reaction patterns and subsequent behavior development.

The behavior repertoire of the newborn infant embraces two large divisions (a) those activities which are generalized body action, non-specific, and, so far as determinable, are not actuated by external stimuli, commonly called "spontaneous activities;" and (b) those reactions which are definitely in response to specific external stimulation. These reactions, although subject to more or less individual variation, are nevertheless made in response to, and can be interpreted in terms of, definitely accountable stimuli. There are unquestionably in the behavior reactions of the newborn infant patterns of response which are specific, and some which are both specific and localized. That is, some specific reactions of newborn infants involve a total body pattern, others are more or less localized to specific members or muscle groups. For example, the Moro reflex is a total body pattern which may be elicited by any number of stimuli, but it is, notwithstanding, a definite reaction pattern. A repetition of the same stimulus elicits essentially the same type of reaction pattern. Reactions to a loud sound, postural adjustments to changes in the plane of the long axis of the body, etc., constitute total body responses. On the other hand, blinking to a tap on the face, tendon and certain cutaneous reflexes are examples of reactions which are relatively localized to limited muscle groups.

In previous studies of infant development little consideration has been given to the relationship of these early reflex responses and subsequent behavior development except in a few isolated cases such as the alleged pathological significance of the persistent "Babinski" or the late appearance of the "Moro." The studies of infant development have tended rather toward the establishment of norms of behavior. Infants are rated in terms of the chronological age at which they attain a certain postural reaction, say sitting or standing, without due credit being given to the process by which that ability was attained. Considerable developmental significance has always been attached to the age at which an infant stands, and walks; and in recent years it has been acclaimed by several writers that acceleration in these traits is suggestive of superior endowment. Gesell (1928:418) says: "The growth characteristic of the infant must prefigure in some ascertainable manner the growth characteristics of maturer years and even behavior traits of those years." During the past decade measures of these "growth characteristics" have sprung into considerable vogue in the form of scales of standardized tests. Most of these scales of measurement for infant development include such items as "holding the head erect in a prone position," "sitting with support," "sitting alone," "standing with support," "standing alone," etc., and are considered to be of developmental importance. It is now well established that many infants only a few hours old, when in a prone or sitting position, will hold the head erect for a few moments; many will momentarily sit with support, stand with support, and take walking or prancing steps. If an infant 10 hours old will support his body weight when held by the fingers or at the axillae, what can be the developmental significance in rating an infant 10 months old who does the same thing? It is admitted that the way the infant 10 months old stands with support is very different from the way the infant 10 hours old stands with support, and it is the business of the infant psychologist to bring these distinctions into relief so that they may be recognized and their significance understood by the less experienced worker.

Neurologically, the difference in the postural responses of the neonate and those of the older infant are probably well defined; the postural responses of the neonate are undoubtedly controlled at a lower, probably segmental level, while those of the older infant are cortically controlled, at least in part.

A study of the reflex behavior of 125 newborn infants by Chaney and McGraw (1932) indicates that specificity of response to many different stimuli has developed at the time an infant is born. Some reactions are more specific than others. The grasping and Moro reflexes are definitely determined behavior patterns functioning with a high degree of specificity at birth.

Since these reactions occur soon or immediately after birth, they are presumably unlearned and reflexive in quality. These early reflex responses, whether localized and specific or total-body reaction patterns, are, it would seem, precursors of, and in some definite way related to, the controlled muscular development of the growing infant. For example, the grasping reflex is a precursor of prehension; the Moro of reaching; primary sitting, standing, and prancing postural responses are forerunners of the assumption of an erect posture and ambulation. A characteristic digital posture of the neonates when the fingers are in extension is a flexion of the distal phalanx of the index finger, a flexion and adduction of the distal phalanx of the thumb, and complete extension of the little finger. This would seem to be a precursor of the prehensile use of the index finger and thumb. Whether these early reactions disappear and cease functioning when cortical control emerges or whether they become an integral part of the cortical reaction is a matter for future investigation. In any event, the process of development from these primary reflex responses to definite muscular control is very gradual and transitory phases are evident.

Outstanding phases through which a course of development passes from reflex to muscular control appears to be as follows:

(1) *Passivity*—that stage when the organism fails entirely to react to the stimulation.

(2) *Reflex*—that stage when the infant responds in a definite pattern to a particular stimulus but the response is of short duration and not under cortical dominance.

(3) *Dyssynergia*—marked by an oscillation of the responding organism— a lack of equilibratory control in sustaining the developing pattern or reverting to a less mature reaction-pattern.

(4) *Inhibition*—marked by an apparent inhibition of a reflex response.

(5) *Control*—denoting muscular control in a given response.

(6) *Synergic Integration*—marked by the control of antagonistic responses so that functionally they are integrated.

These stages are well illustrated in the assumption of an erect posture. At the time of birth some infants are decidedly passive and show little resistance to the pull of gravity; others have already reached the reflex stage and momentarily support their body weight when given a little assistance. As the baby develops in control, marked dyssynergia is noted. He gains control of the movements of his head and neck before he has control of the trunk and lower extremities. Finally, when he has developed sufficiently to stand alone momentarily, dyssynergia is less frequent, and this development of standing has an inhibitive effect upon his "dropping" down to a sitting posture before he is able to *sit down* cautiously. There is, so to speak, an inhibition of an opposing function. A little later he gains control of standing and sitting and the process of getting up and its antagonist, getting down, become integrated and under the complete control of the child.

So it would seem that the true definition of a reflex is not a question of simplicity of responses or limited synaptic connections; nor is it a matter of non-modifiability or localization or precision of the reaction pattern. The term reflex should include all those reactions of the newborn having a characteristic pattern of reaction to known external stimuli, whether the reactions are total body patterns or localized and specific. Then the question arises as to the importance of the primary reflex patterns in the process of controlled muscular development.

CONCLUSION

So far as the behavior of the newborn and growing infant is concerned, there is evidence supporting antagonistic theories on the nature of neuromuscular growth. Both localized and total-body reaction patterns are present in the behavior of the newborn infant. These reaction patterns, although distinct in type of response, are subject to individual variability and modifiability. So it would seem that the old psychological definition of a reflex is in line for revision, so as to include all those reactions of the newborn (whether localized or a total-body response) having a discernible characteristic pattern. Certainly these primary reflex responses of the newborn bear some ascertainable relation to the controlled muscular behavior of the older infant, young child, and adult. Just what the nature of that relation may be is a matter for future investigation. There is as yet no conclusive evidence as to whether or not behavior metamorphosis as observed in the growing infant is a process of aggregating synaptic connections or a process of "individuation" or inhibition of accessory responses. Certainly, the change from reflex or subcortical to controlled or cortical behavior in the infant is very gradual. There is no evidence of a sudden shift from one type of reaction to another, hence no indication of a sudden maturation of function. As a matter of fact, the aspect of dyssynergia accompanying the emergence of any new postural response in the developing infant is highly suggestive of trial-and-error learning. Learning and maturation are not two distinct processes but are two aspects of the same process. To attribute behavior growth in infants more to one than to the other is, therefore, unwarranted. There comes a time, however, due to the ripening of neural structures when systematic practice of a given function will have greater effect upon improvement of performance than it would at any other time. To attempt systematic practice of a particular function before the neural structures have obtained a degree of maturation is ineffectual. The performance of a given function does not indicate that structural maturation is completed. To fail in the practice of a given function when the time is ripe curtails improvement in not only the overt performance but probably lessens future maturing of the particular structures involved.

9

Behavior Development

The essence of childhood is growth. Growth of children is so ubiquitous that for thousands and thousands of years it failed to arouse curiosity or stimulate inquiry as to the nature of its development.

Recently that unconcern has given way before a wave of intense interest in the development of childhood. The technical scientific interest of the experimentalist, the practical interest of the clinician, and the curious eager concern of the layman are well represented in the vast array of publications on the subject. Elaborate research centers have been established for the sole purpose of studying the development of the child, and the physical, mental, emotional, motor, personality, and behavior development of the child have received both intensive and extensive analysis during the past quarter of a century. The infant and young child are now recognized as useful laboratory animals. They have served as subjects for innumerable types of studies: studies in the development of skeletal and muscular structures, the development of lymphatic systems, the nature of perceptions and sensations, and growth in social behavior.

Wherever children are used as subjects in investigations the study is automatically listed as one in "child development." The term seems to suggest that child development is unique, or different from any other kind of development. Emphasis upon development during a particular chronological period—childhood—implies furthermore that growth during childhood is in some way distinctive from growth at any other time. Classification of the literature pertaining to child development in terms of chronological periods—for example, fetal development, infant development, development of the pre-school, the pubescent, and the adolescent child—lends support to the implication that development during each chronological period is distinct in itself.

Reprinted from Myrtle B. McGraw, *Growth: A Study of Johnny and Jimmy*, Copyright, © 1935, renewed 1963, pp. 3–21. Reprinted by permission of Prentice Hall, Englewood Cliffs, New Jersey. Reprinted by permission of Mitzi Wertheim.

161

Although the literature is replete with studies of physical, mental, emotional, motor, and personality development at these several chronological periods, there has at the same time been considerable stress upon the need of studying "the whole child." So far as the writer is able to perceive, a study of the whole child has meant primarily an aggregation of these items. The child has been considered more or less as a composition of compartments. When the status of each compartment is determined for each chronological period, and the aggregation of these various ratings computed, it has been assumed that the development of the child would thereby be revealed. That is, inventories of the developmental status at these several chronological periods, when pieced together, would presumably yield the story of development during childhood.

Not only has there been a tendency to break childhood up into convenient chronological periods, but the method of studying development during a given period has been largely that of cross-sectional analysis. This tendency persisted despite the loud praise which has been awarded to the genetic method. With the exception of a few instances those studies which purport to be genetic in method are in fact little more than inventories of the same child-subjects at different chronological periods. While these inventories represent increments of growth and are highly useful for certain purposes, they do not afford an analysis or understanding of the *process of development*. The student of development in behavior must focus attention upon the movements which give rise to an organized behavior pattern, as well as the chronological sequence in which behavior patterns reach maturity.

Wherever there is development there must be a period of preparation, consummation, and decline. It seems that many apparently controversial theories of behavior development are not really controversial but appear to be so because they have been derived from different stages of maturation. The experimentalist has too often undertaken to establish principles of behavior development from data which represent only a phase or a section of the developing process. The tendency to study the child in cross-sectional periods has undoubtedly provoked polemics on the nature of behavior and development, polemics which might not have arisen had the principle been derived from the entire developmental process rather than from a section thereof.

It would be well to consider here some of the more recent theories of behavior, especially as they have reference to the behavior of infancy and childhood. The development of a theory is in principle fundamentally akin to the development of behavior. If it were possible to recapitulate the processes involved in the development of theories, and concepts of behavioral growth, we should probably find that the stages and characteristics of such growth in knowledge are similar to those of the infant in learning a new performance such as, for example erect locomotion. In the beginning he is capable of reflex stepping movements only, and these he makes with little if any awareness or control on his part. As he becomes actively engaged in the developmental process he manifests a tendency to swing from one extreme to another, as, for instance, from an unusually narrow base to an extremely wide one, a tendency to exaggerate or overwork any new aspect of development, a decided dyssynergia, and a display of excess or overflow of activity. Gradually the excess activity is eliminated, the exaggerations are narrowed down, and an ever increasing specificity of pattern is by degrees manifested until a well-coordinated and integrated gait is achieved. Analogously, for generation upon generation whatever body of knowledge adults had concerning the behavior development of children was obtained through no direct conscious effort to study child development. Fortunately, children continued to grow despite the general unconcern over their manner of growth, and fortunately adults actually learned a good deal about development although they were not deliberately trying to do so. We have now, however, arrived at the stage of making conscious effort to determine principles of behavioral growth. Sometimes it even seems that too much effort is being exerted. Too much motion is being wasted without actually stepping forward. An idea or new theory emerges and shows the characteristic tendency to exaggerations in its claims; then in time the growth of that particular theory is checked as a new, counteracting one crops out. The new one in turn also becomes excessively ambitious and exaggerated. Finally the essence of the two may become crystallized into an unchallengeable truth, a specific, well-determined principle which can stand alone and walk through the universe of ideas fullgrown. The study of infant behavior development is young, and such theoretical fluctuations are the signs of youth.

THEORIES OF BEHAVIOR DEVELOPMENT

One of the earliest theories of development having reference to infant behavior did not arise from studies of infants, nor was it concerned primarily with development. It essayed to explain behavior. Early psychologists, being inclined to think of all behavior as either inherited or acquired, were accustomed to classify behavior into reflexes, instincts, and acquired traits. According to this theory an infant was born equipped with a system of reflexes and instincts which were inherent in the germ plasm and presumably not subject to appreciable modification through experience. It was customary for the psychology textbooks at that time to include long lists of reflexes and instincts. All other types of behavior were considered as acquired traits. Although some instincts were delayed in making their appearance until the latter part of childhood, experience or environmental factors were of no consequence in determining their appearance. The disconcerting feature of this classification of behavior was that the list of instincts became very elastic. No two textbooks agreed, so that the term came to embrace a large share of human behavior. Obviously, the usefulness of such a classification of behavior was lost on account of its elasticity.

This theory implied that the process by which reflexes and instincts were developed was distinctly different from the process by which acquired traits were developed. Reflexes and instincts, being passed from generation to generation through the chromosomal genes, presumably developed in one way whereas individual attainments developed in another. Having attributed reflexes and instincts to organic inheritance, having placed their locus of origin somewhere within the genes, the psychologists made no further attempt to explain the way in which they did develop. The burden of explanation as to the process by which reflexes and instincts developed rested with the biologists and geneticists. The psychologists were content to recognize that they did exist and that their manner of development was different from that of acquired traits. Acquired traits were primarily a product of experience. In the development of individual attainments environmental factors were paramount. This theory might have lived longer and served a more useful purpose if the psychologists could have agreed upon a classification of instincts. Since they could not, the theory had to give way to another.

It is in line with the typical process of development that the succeeding theory should have proclaimed the significance of environmental factors in the determination of behavior and development, since the instinct theory had attributed so much weight to original endowment. It is the contention of the early Behaviorists that the infant is born into the world equipped with a limited number of reflexes, or simple reaction mechanisms. By a process of *conditioning* these simple reflexes are concatenated into reactions of increasing complexity. In the conditioning process the environmental factors are of prime importance. Every one is familiar with the celebrated statement of Dr. Watson (1930) the chief protagonist of Behaviorism, in which he proclaims the relative influence of environment in the development of the child. "Give me," he says, "a dozen healthy infants, well-formed, and my own specified world to bring them up in and I'll guarantee to take any one at random and train him to become any type of specialist I might select—doctor, lawyer, artist, merchant, chief, and, yes, even beggar-man and thief." While many Behaviorists may not go as far as Dr. Watson in stressing environmental factors as opposed to innate endowment, there has been an undeniable tendency of this school to emphasize the extrinsic influences in behavior and to accept the conditioning theory as adequate explanation of behavior development. The term *conditioning* has therefore expanded far beyond its original meaning, so that it is now often used loosely to indicate any associational behavior. However, since the Behaviorists admit that the infant is endowed with a few simple reaction mechanisms, they imply that the way in which these few reflexes are developed is in some manner different from the way in which conditioned reflexes or learned activities are developed. As long as they refer to *learned* and *unlearned* behavior, they suggest two different processes of behavior development. So far as a theory of development is concerned, the difference between the Behaviorists' point of view and that of the earlier reflex-instinct theory is primarily one of emphasis. The Behaviorists, restricting innate endowment to the minimum, emphasize environmental factors.

It is not surprising that there was a reaction against this highly mechanistic doctrine of development. In the field of infant behavior Gesell (1933) has most actively voiced opposition to the environmentalists and revived the significance of innate endowment. He has dressed it in a coat of a different color and called it "maturation." He defines maturation as "the intrinsic component of development (or of growth) which determines the primary morphogenesis and variabilities of the life-cycle" (Gesell 1933:231). Gesell has repeatedly asserted the priority of the ripening process of neural structures preparatory to function. The neural ripening, he contends, is uninfluenced by environmental factors. He claims,

> The nervous system grows according to its own intrinsic pattern and thereby establishes the primary forms of behavior. These forms are not determined by stimulation from the outside world. Experience has nothing specifically to do with them ... the extreme versions of the environmentalists and conditioning theories suffer because they explain too much. They suggest that the individual is fabricated out of the conditioning patterns. They do not give due recognition to the inner checks which set metes and bounds to the area of conditioning and which happily prevent abnormal and grotesque consequences which the theories themselves would make too easily possible (Gesell 1933:232).

In drawing the distinction between *maturation* and *learning* Gesell implies that there are two distinct processes of development and that development during infancy is chiefly of the maturational type. A large portion of the studies of infant behavior which have emanated from the Yale Psycho-Clinic lend support to his thesis that behavior development during infancy is relatively uninfluenced by individual experience.

Another theory of development in opposition to the conditioning theory has been concerned not so much with intrinsic and extrinsic factors in behavior as with the mode of development. While this theory did not originate from studies of infant behavior, many of the more recent investigations concerning infancy have been interpreted in terms of *individuation*. The theory of individuation emanated from a laboratory in the Wistar Institute where Coghill has been for many years patiently and unostentatiously watching the first embryonic movements of *amblystoma* [salamander]. He has followed these movements systematically through every stage of development until the animal and its behavior reach maturity. Coghill has correlated the developmental changes in the somatic behavior of *amblystoma* with progressive changes in the structure of the nervous system until he is convinced that the process of development is not one of synthesis or compilation of increasingly complex patterns, but one of analysis or individuation. He explains:

> It has been generally assumed that larger patterns of behavior are formed in the development of the individual by the integration or synthesis of smaller or elementary patterns. Conversely, the thesis of this paper is that smaller or partial patterns arise by a process of individuation or analysis within a larger pattern and that the primary pattern is a total reaction which normally expands from the first as an integrated process. ... Conditioned reflexes, which are generally accepted as the units out of which behavior develops synthetically in post-natal life, are generally regarded as differing in their genesis from the unconditioned reflexes of fetal life. It is probable, however, that this hypothesis is not well founded. Pavlov's dogs which condition their reflexes in an orderly way do so by progressive reduction of the field of adequate stimulation for the particular reflex and by progressive restriction of the motor field of action. These characteristics the conditioned reflex has in common with the unconditioned reflex. The conditioned reflex, like the unconditioned, is acquired by analysis of a total pattern which under normal conditions is from the beginning perfectly integrated (Coghill 1930a:431; 434–435).

Coghill sounded a note in advance when he ascribed the same fundamental processes to the development of localized discrete reflexes as to conditioned reflexes and learning by trial and error.

> The conditioned reflex is conventionally regarded as differing essentially from the unconditioned reflex, but this is contradicted by evidence drawn from the development of behavior. Like unconditioned reflexes, the conditioned reaction emerges on the motor side from a field of general activity and on the sensory side the specific stimulus emerges from a general or wide zone of adequate stimulation. ... On the motor side of the conditioning, reaction is at first general, approximately a total reaction, at least of a postural nature, and only later does it become specific. Conditioning of reactions accordingly is accomplished by restriction (narrowing) of the zone of adequate stimulation and concomitant restriction in the field of action. The primary structural basis for this is in the mechanisms of the total pattern. The same interpretation can be applied also to learning by trial and error. The immediate response of an animal when placed in an utterly strange problem box or cage is general activity; and the stimulus to this action appears to be the situation as a whole. But within this general situation there eventually emerge particular features as relatively localized stimuli, and concomitantly out of the general field of action there emerges the particular act that is appropriate to the situation. Here again the mechanism of the total action pattern is the key to the process of learning (Coghill 1930b:640).

Coghill has presented a theory of development which applies to any level of activity. He has shown that infracortical and cortical behavior-patterns manifest the same underlying principles in their course of development. Furthermore he has emphasized that the principles of growth are general whether the factors involved are organic or functional. That is, the principle of individuation applies in the structural development of the nervous system and in the overt behavior of the animal. Studies in experimental embryology and genetic psychology have been knit by Coghill's suggestion of a fundamental principle of development into a closer relationship than has ever before been conceived.

It is sometimes claimed that the Gestalt school of psychology is the one whose theories on behavior most closely parallel the work of Coghill. While the Gestaltists have not been concerned primarily with development and their theses do not constitute a theory of development, the general contentions of this school and their supporting experimentation oppose the notion that complex "wholes" are constructed out of simple elementary processes. To this extent they may be identified with the theory of individuation. Among the students of infant behavior Irwin (1932) has most enthusiastically espoused the theory of individuation in depicting growth of specific reactions out of a general mass matrix as indicated in the activities of the newborn infant.

But it is the experimental embryologists and not the psychologists who deserve credit for formulating the most adequate theory of behavior development. It is they who are revealing the process of morphogenesis, and it is they who are bringing the most convincing experimental evidence to bear upon an evaluation of intrinsic and extrinsic factors in the process of growth. Let us examine some of the principles which have been educed from studies of embryological development.

In many ways development as manifest in the early metamorphosis of the germ cell is extraordinarily similar in principle to that shown in the development of behavior in the infant and young child. Growth of the germ cell shows a primary stage of rapid development immediately after fertilization. This period of rapid development is followed by a static period when smaller and fewer changes take place. It has already been established that early in the period of cleavage particular cells or groups of cells become destined to develop into particular organs or systems. Even before the gastrula stage it is possible to determine the specific region of the embryo—barring mishap—which will develop into the epidermis, the notochord, or the mesoderm. To say that they are destined is a little misleading, for that would seem to imply that there is no other course for them. As a matter of fact, they are destined to develop into the particular organs *provided* there is no interference with their customary course of development. But there is a period of "indeterminateness," when their fate, though suggested by the position of the particular cells, is not finally settled.

Although it is possible to ascertain definitely the organs into which these cells are likely to develop, there is a period of indeterminateness in which it is possible, mechanically or otherwise, to change their course of development completely, so that for instance the germ region which would ordinarily give rise to viscera may develop into epidermis. It has been demonstrated experimentally that it is possible to implant a bit of embryonic region which would ordinarily give rise to the development of bone tissue in the region which is destined to become epidermis, and that the implant, assimilating its new surroundings, will develop into perfectly normal epidermal tissue. But the most interesting thing about it is that there is a critical period when the transplantation must occur in order to bring about these results. At a later stage such a transposition of embryonic cells will have strikingly different results. The cells will then develop in accordance with their origin and not in accordance with their position. That is, transplanted presumptive embryonic bone tissue will, at a later stage, develop into bone even if it has been placed in the environment of presumptive epidermis. At this later period these cells have reached a stage of specificity, their growth destiny is sufficiently determined, so that they are resistant to the influences of their new environment. To recapitulate, embryonic growth reveals a period of indeterminateness when the destiny of particular cell groups may be indicated to the extent of organ rudiments, but not so fixed that their development cannot be altered by placing these rudiments in a new environment. Then there follows a period of greater determination in growth when the cells, true to their early influences, resist modification even when placed in a new environment.

It has also been attested that once a region of cells has differentiated or attained specificity there pass out from it influences which govern the destiny or determination of other embryonic cell regions. That is, the differentiation of one group of cells constitutes an organization center which exercises an influence upon the developmental fate of other cell regions. For example, the development of the optic cup is indispensable to the lens. If the rudiment of the optic cup is extirpated before the formation of the lens, then the lens never becomes fully differentiated. Furthermore, if a bit of epidermis which normally would develop into a lens is replaced at an appropriate time by a piece of epidermis from the trunk, then the lens is formed from this foreign transplant. If the trunk epidermis is transplanted at a later stage the lens will not form. And if the primary optic cup is transplanted at an appropriate time under any other part of epidermis it will often stimulate the development of a lens from the epidermis in this foreign position. The development of the lens is contingent upon the relationship between optic cup and epidermal tissue at a favorable stage of immaturity.

From the above illustrations it is apparent that all regions of the embryonic cells are not developing at the same time nor at the same rate. There exist in the same germ side by side parts which are fully determined, others which are rudimentary but still somewhat indeterminate, and still others which have not even begun to differentiate or evidence signs of organ development. Not only do different regions of embryonic cells begin their development at different times, but the periods of indetermination or plasticity are of unequal lengths. Since different cell groups are in different stages of development at any particular time it would be impossible to express the degree of development of a particular embryo in chronological terms from the moment of fertilization. Both position and time, it must be remembered, are important in the development of the embryo.

The process of embryonic development is from the general to the particular, but the progress from one phase to another is extraordinarily gradual. In emphasizing the gradual, step-by-step process of development from an undifferentiated to a specific and determined stage, the writer cannot do better than to quote Durken (1932:121–122), from whose scholarly work she has drawn materially in this presentation of embryological development.

> This determination comes about by degrees, earlier or later according to the kind of animal in question, so that an indeterminate condition gradually gives place to a determinate one. Further, since all the parts are not determined at the same time there arise in the germ local and spatial differences in the state of determination. But determination in itself is a gradual process. Not only does it lead little by little from the general to the particular, but the rigidly determined condition is connected by intermediate stages with the indeterminate condition, beginning with the first appearance of determination—when it is still indefinite—and proceeding to its final complete establishment. In order to distinguish between these degrees of determination we may use for the introductory phase the term *institution*, and for the final phase—in which ontogeny no longer remains uncertain—the term *destination*.

Not only during the stages of cleavage and early embryonic development, but during the fetal stage—in fact throughout the entire ontogenetic cycle—these aspects of organic development are evident. It is established not only that all organs of the developing embryo and fetus are not formed at the same time, that the formation of one organ may have an influence in the determination of another or subsequently developing organ, but also that the new organ, when it arrives at its period of rapid growth, may have an arresting or inhibiting effect upon the development of the organ previously developing. There is, it appears, a sort of growth competition between the various organs during their periods of development. This competitive relationship operates to gain a proper functional balance. It is important that each organ should have its period of rapid growth at the normal time, for otherwise the formation of a new organ may have an inhibiting effect before the former has reached its full development. This would result in a stunted organ and thus destroy the functional balance of the organ interrelationship.

The theory of epigenesis is therefore the first to suggest that development is a *process* having fundamental principles which inhere despite the nature of the organism which is growing. Development is a process having a period of origin, incubation, consummation, and decline. Therefore in order to study development it is necessary to initiate study at the inception of the growing organism or function and to follow the process through all the succeeding changes until it attains maturity.

PURPOSES OF PRESENT INVESTIGATION

The present investigation purports to analyze the process of development as it is manifest in the behavior of the growing infant and young child. When one proposes to study behavior development of the human infant he is beset with dire problems, not the least of which is the determination of the origin of the organism to be studied. Since in studying the process of development it is necessary to begin study at the time of origin, the locus of origin must be established. To establish the point of origin of an individual child is a problem which transcends experimental procedures. Does the new individual begin with ovulation, fertilization, or respiration? If conception is arbitrarily selected as the origin of the individual, then it must be admitted that considerable preparatory development has gone forward during ovulation, before fertilization and cleavage occur. Furthermore, even if conception is accepted as the most logical period for designating the beginning of the individual, the difficulties involved in initiating study of human behavior during fetal development are almost insurmountable. While no one would question that cellular cleavage is a phase of behavior development, it does not easily lend itself to experimental observation in the human embryo. So for all practical purposes, in this country at any rate, human fetuses are not often available as experimental subjects until respiration is established. It is obvious, therefore, that the experimentalist cannot initiate analysis of all human behavior at the time of origin.

Moreover there are other problems involved in considering the individual child as a unit of investigation in studies of behavior growth. Since the behavior of the human infant has attained a high degree of complexity and organization by the time the baby becomes available to the experimentalist, a mere listing of all types of behavior of which he is capable is impractical, if not impossible. A systematic genetic study of his total behavior-repertoire is even less feasible. Even if it were possible to record all the behavior of which the average infant at a particular age is capable and then all the behavior of which a slightly older infant is capable, such listings would not reveal the process of growth. Inventories of behavior even if arranged in chronological series do not reveal the movements which give rise to a particular type of behavior. It is an interesting fact that the average baby sits alone at eight months and that he walks at fifteen, but these facts constitute neither a description nor an explanation of the process of growth. Walking is the *end* result of a process of *development*, whereas the study of development in behavior must be concerned with the process by which the end result was attained. In order to understand the development of behavior the genetic psychologist must ascertain the anlages out of which a particular behavior pattern arises and the phases through which it passes before attaining maturity as adequately and accurately as the embryologist determines the cells or groups of cells and the conditions in which those cells give rise to a particular organ.

Since it is impractical to study all the behavior-patterns of the growing child from their origin to maturity or fixity, a possible method consists of the selection of only a few characteristic behavior activities in order to follow the progressive changes manifested in the course of development of these patterns. For convenience we shall use the term *behavior-course* to designate an organized function, such as erect locomotion, during the process of development. *Behavior-pattern* refers to the terminal form of activity, the end result. We prefer the term *course* as representative of the growing period since the word suggests movement and change as against the boundaries and fixity which the term *pattern* connotes. There are many different aspects and phases involved in a behavior-course before it develops into the stability of a pattern. It is these changes and phases which are of moment to an analysis of behavior development.

Not only is it impractical to study all the behavior-patterns within the capacity of a given individual but it is equally impractical to analyze all the aspects or ingredients which enter into the development of a behavior-pattern. In the present investigation we have elected to restrict our analysis to overt-somatic changes which can be observed in a behavior-course as manifest in the activities of the growing infant and young child.

In proposing to make analyses of the development of particular behavior-patterns we must avoid the implication that a pattern of behavior can be isolated from the individual and studied as a unit. It is at all times an integral part of the total behavior of the individual and is constantly influenced by intrinsic and extrinsic factors which enter into the development process. Although the behavior-pattern is an integral part of the individual, from which it cannot be dissected, it nevertheless retains its own identity at all times. Fortunately, dissection is not necessary in order to analyze its course of growth. It is possible to focus attention upon a particular behavior-pattern, or certain aspects of the pattern, in order to follow its course of development from the time of origin until it attains maturity or fixity.

A second purpose of the present investigation was to study the *modification* of behavior development during infancy. Behavior development is a process of interchange of energies within an organism and energies within its environment. In order to bring about a modification of development it is necessary to alter the energy relationship between the organism and its environment. Actually, wherever there is development there is a constant change of energy relationship. But in this treatise we are using the term modification to indicate that the course of development is being altered from that which prevails under ordinary or customary conditions. Thus considered, modification may be either in the nature of (1) acceleration in the *rate* of growth, (2) retardation in the rate of growth, or (3) alteration or interruption of the serial sequence in the ordinary phases of development. Modification of development in the sense in which we have used it is not merely change but change from the usual to the unusual.

The interrelationship of intrinsic and extrinsic factors in development may be altered by an addition to or a subtraction from either the organic or the environmental factors which enter into the growth process. In general it is easier experimentally to manipulate the environmental factors than the organic. The environmental factors alone which influence a given growth process are multitudinous. Therefore, in order to study the effect of one or a particular environmental influence, it is necessary to select the particular extrinsic factor and increase or decrease its action upon the growing organism. We have seen how embryologists have determined the influence of heat, humidity, and light upon the growth of particular organs or organisms. Comparably, in a study of human behavior the experimental factor may be that of temperature, light, and so on. But in the present analysis we have endeavored to ascertain the effect of exercise or use upon the growth of particular action-systems.

We elected to study the influence of increased or restricted use of an activity upon its course of growth. Since we were using infants as subjects, study was further limited to an evaluation of the influence of exercise upon the development of particular behavior-patterns during the period of infancy.

INTRINSIC AND EXTRINSIC FACTORS

It has been said that behavior development is a process of interaction of organic and environmental factors. But actually the distinction between internal and external is an arbitrary one used solely for the purpose of convenience. That which is external or environmental from one aspect is internal from another. At one time the same factor is extrinsic and at another it is intrinsic. In popular parlance, and in much of the psychological literature, it is customary to speak of those things as internal which are contained within the epidermis. Yet the body is full of foreign substances at all times. Food at one stage of digestion is external and at another internal. A disease germ within the body is medically considered as a foreign substance within the body. In the study of behavior the same disease germ might be considered an intrinsic factor. Epidermis might serve as an adequate demarcation between intrinsic and extrinsic for one purpose but be completely out of bounds for another. In order to study the development of an individual it is necessary to decide arbitrarily where the individual ends and the environment begins. Such demarcations are defined by the experimentalist for his own convenience in analysis.

Likewise in a study of the growth of a behavior-pattern it is necessary arbitrarily to define the boundaries within which the behavior-action will be analyzed. The consecutive serial changes which occur within the limits defined are an indication of development. Of course, the whole organism is involved in any behavior activity, but there are centers or concentrates of action which constitute a pattern of behaving. It is no more unreasonable to focus attention upon these centers of action within an individual in order to observe their course of development than it is to concentrate upon the growth of an organ within the body or to fixate upon the development of an individual within a society.

While the whole organism is involved in any action of the individual, behavior is nevertheless the primary concern of the nervous system. Morphological changes of the nerve cells, cellular lamination, and myelinization of nerve fibers are just as much a matter of behavior development as are changes in the overt behavior-patterns of walking, fighting, or writing a book. In this investigation we have not been able to cover all of the aspects of development which enter into the growth of even one behavior-pattern, to say nothing of all the actions which comprise the development of an individual. Systematic analyses of the changes as indicated in overt-somatic activity in the development of a behavior-pattern from the time of origin until it attains maturity constitute both a feasible and a revealing method of studying the process of growth. We have, therefore, restricted our analysis to the development of a few select behavior-patterns as they can be seen in the somatic activity of the infant and young child.

Since in the human infant the inception of many activities has occurred before the birth of the baby, we have found it convenient to classify the behavior-patterns of the infant in terms of the degree of fixity which the behavior-course has achieved at birth or at the time the subject becomes experimentally available. While the infant is young in ontogeny he represents the terminus of a long phylogenetic series; therefore a classification of infant behavior-patterns in terms of both phylogenetic and ontogenetic significance is necessary in order to study growth of the activity.

CLASSIFICATION OF INFANT BEHAVIOR

There are certain types of behavior-patterns which have *in utero* attained a high degree of specificity. By the time the infant is born he is able to exercise these functions about as efficiently, and in essentially the same manner, as he does during adulthood. They are subject to very little modification during post-natal development, insofar as form or pattern of reaction is concerned, through post-natal external influence. The newborn baby yawns, coughs, sneezes, hiccoughs, stretches, cries, and sleeps about as well immediately after birth as he will ever be able to do. To be sure these patterns of behavior may be slightly modified or controlled as the child grows older, but it is impossible to take the pattern of coughing and make something completely different out of it, any more than the embryologists can, during the later stages of development, take a kidney or intestine and make it grow into epidermis. During the early stages of embryological development, as stated before, such an interchange is quite feasible by transplantation of germ regions. Whether it is possible to catch the presumptive coughing pattern at the time of origin, or before it has attained specificity or determination, and by altering influential factors, develop the presumptive pattern into something quite unlike coughing is not known. Traits such as these which are fully developed at birth have a long period of functional maturity, since they exist throughout the life of the individual. They are so characteristic of the behavior of mankind that any major deviations in their functioning are considered pathological.

A second group of behavior-patterns manifested in the activities of the newborn have also attained a high degree of specificity during pre-natal development, although they do not in the human infant retain their usefulness. In the absence of a better term these behavior-patterns have been called atavisms because of their apparent phylogenetic significance. To give them this classification is in a way objectionable since it involves an interpretation of their origin and development which our present data do not afford. Some behavior-patterns of the newborn appear to be residual rudiments of activities which, earlier in the phylogenetic scale, were functionally useful. In the infant they appear to be reminiscent of a primordial function which in ontogeny is comparatively inessential to the well-being of the individual. The distinctive feature about them is that the overt action-pattern tends toward a diminution or weakening as the child grows older. The persistence of these patterns or their recurrence after a given time is considered pathological. An outstanding example of behavior reactions of this type is the Moro reflex. It is sometimes known as the embrace or body startle reflex of the newborn. Any number of stimuli will elicit such a reaction in the newborn baby, but one of the most common methods is to slap upon the bed or table on which the infant is lying. The theory is that it is a hangover of the clinging reflex common to primates. The young simian immediately after birth will clutch the mother's fur on the abdomen and cling to her as she jumps about the cage. The reaction of the human infant is incomplete and merely suggestive of the primitive reaction. The developmental sequence of this behavior pattern is reported in detail in Chapter 3 [of *Growth*].

A similar atavistic behavior-pattern of the newborn infant is sometimes called the Darwinian or suspension grasp reflex. The newborn infant will grasp so tightly on a small rod, if it is placed in his hand, that he is able to raise his body above the surface on which he is lying and remain suspended for a few seconds by his grip on this rod. The progressive phases in the development of this trait and its pertinence to the *process* of growth will be discussed in a later chapter. Crawling movements, stepping movements, and swimming movements are similar residual patterns of a function of considerable phylogenetic import, but patterns whose manifestations in the human infant appear to be merely residual of their earlier usefulness. In the newborn infant, they are approaching the stage of decline, so that development is indicated by a gradual diminution of overt action. It is the evanescence and not the expansion of activities of this type which indicates development.

Behavior traits of a third order, which have not definitely emerged in the activities of the newborn, are nevertheless of phylogenetic significance and are indispensable to normal human functioning. Traits which would fall into this grouping are reaching-prehension, creeping, the assumption of a sitting posture, and erect locomotion. Certainly it is determined long before the baby makes his first step that, barring mishap, he should ultimately stand on two feet and walk erect. The infant who does not in time acquire erect locomotion is pathological. Rickets or even the prevention of exercise in this activity may within a measure delay the development of certain aspects of the pattern or alter to a slight degree the form of its development, but if any external influence is sufficient to prevent the child from walking on two feet then he becomes a subject for pathology. These traits thus stand in unquestioned distinction from specific skills or abilities which may or may not be acquired by an individual.

Thus the fourth classification of behavior-patterns would include the vast number of skills and abilities the development of which is not indispensable to the life and normal functioning of the individual. It is essential that all normal children learn to walk erect, but it is not essential to normal life that they learn to ride a bicycle or to roller-skate.

It is, however, extremely pertinent for an understanding of the process of development that we ascertain whether or not the same processes and principles are involved in the acrobat's walking a tight-rope as in the infant's acquisition of erect locomotion, or in the development of organs, or cellular cleavage.

✤ 10 ✤

General Principles of Growth

Since development is a factor common to all living matter, it is not unreasonable to assume that there are fundamental principles of development inherent in the process of growth despite the nature or character of the growing organism. We speak of the growth of a cell, a plant, a tumor, a disease, a culture, a society, an idea, or a behavior-pattern. Do we mean the same thing when we speak of growth in such diverse connections? If development is a process having general identifying characteristics which transcend the particulars wherein the growth occurs, then an adequate theory of development must embrace the process involved in cellular cleavage, organ differentiation, and simple and complex behavior activities.

The psychologists have failed to arrive at an all-embracing theory of development in behavior because too often their subjects, whether plants, lower animals, or humans, have been adults, or, if the subjects were not adults, then the particular trait under observation was mature or well determined at the time study of it was initiated. The limitations of the conditioning theory, for example, lie in the fact that it is constructed out of reaction patterns which are mature. Under such conditions growth would be a matter of associations or rearrangements of patterns already developed, whereas development really involves the *emergence* of something new—a way of behaving in which that particular individual has never behaved before. The ability to secrete saliva is a fully developed reaction pattern in the dog. By conditioning he learns to secrete saliva at the ringing of a bell as well as at the sight of food. One type of stimulus has been substituted for another, but the dog's manner of reaction is the same as it was before the conditioning.

The baby can be conditioned to cry at the sight of a rabbit, though it is presupposed that he had the ability to cry long before he saw the rabbit. Again, a newborn baby is equipped with a well-developed sucking mecha-

Reprinted from Myrtle B. McGraw, *Growth: A Study of Johnny and Jimmy,* Copyright, © 1935, renewed 1963, pp. 300–312. Reprinted by permission of Prentice Hall, Englewood Cliffs, New Jersey. Reprinted by permission of Mitzi Wertheim.

nism. By ringing a buzzer every time a bottle is presented to the infant it is possible to condition him so that the mere ringing of the buzzer will set up the sucking response, though there is no nipple to suck on. So far as the infant's overt response is concerned, nothing new has been created in his behavior-repertoire. But one could ring a buzzer a thousand times over simultaneously with the presentation of the bottle and yet not get a three or four months old infant so *conditioned* that he would take the initiative of deliberately ringing the buzzer in order to have the bottle brought to him. The problem in development is to ascertain what has occurred to make it possible for the older child to take such initiative, whereas no amount of conditioning was sufficient to develop it in the growing baby.

When the inseminated egg begins to divide, it is behaving in a way that particular egg has never behaved before; and when the baby first pulls himself up and stands alone, he is behaving in a way that particular baby has never behaved before. These changes in behavior are the phenomena characteristic of development.

The sort of development that means a mere *substitution* of stimulus or response is different from that type of growth which involves the creation or emergence of a new type of behavior; new, that is, in so far as the particular individual is concerned. It is, therefore, necessary to distinguish between situations which involve merely a substitution of stimuli or responses and those which actually create a new pattern in the overt-action-system. An adequate theory of behavior development will embrace the process of cellular cleavage; the infant's learning to walk, the ten-year-old's learning to ride a bicycle, and the college student's solving a problem in calculus. An adequate theory of development must embrace the emergence of new behavior-patterns. But one does not study development as such, one studies the development of something. One may study the development of a plant, of ossification, of a social order, or of an infant.

In the present investigation we have chosen to study the development of a few select behavior-patterns as they are manifest in the growing infant and young child. Of course a behavior-pattern cannot for purposes of study be dissected from the infant of which it is an integral part, nor does it function as an isolated unit distinct from and uninfluenced by other sections of the child. Nevertheless, we have shown, we believe, that it is possible to focus attention upon the development of a particular behavior pattern, as it is seen within the individual, in the same way in which one might focus attention upon an individual child as a specialized nodal point of the environment in which he lives. In fact it is scarcely more difficult to establish the origin and continuity of a behavior-pattern within an individual than it is to establish the origin and continuity of an individual within a society. The individual represents a nodal point of specialization within an environment; correspondingly, behavior-patterns are nodal points of concentration of activity within the total action system of the individual's behavior.

In any investigation it is necessary for the purposes of analysis to set arbitrary bounds and restrictions to the material upon which attention will be focused. Therefore in studying the development of behavior-patterns we have deliberately limited our analysis of those growth changes which can be observed in the somatic movements of the subject. Of course, morphological changes of the nerve cells, cellular lamination, and myelinization of nerve fibers are just as much a matter of behavior development as are the sequential changes in the achievement of walking, fighting, or learning to read. But unfortunately we are not able to analyze all the factors which enter into the development of an action-system. Since, however, the development of behavior as indicated merely in the overt-behavior patterns herein described enhances the advancing information on the nature of growth *per se,* as well as pointing out the phases of growth in these particulars, this detailed account of the development of select behavior-patterns gains a significance beyond the factual details.

When the course of a behavior-pattern is followed systematically, the appropriateness of the term *pattern* becomes more evident. The growing of a behavior-pattern is likened to a design in the process of being woven, composed as it is of various colored threads. All of the threads do not move forward at the same time nor at the same rate. The weaver picks up the gold thread and weaves it back and forth, though at the same time steadily forward. Then he drops it in order to bring the blue thread forward a distance, until finally the two become united to make the pattern complete. The design is contingent upon the interrelation of the various threads. It is not the summation of the blue and gold threads but their position with respect to each other and to the piece as a whole which determines the design. To unravel one of the threads would mean destruction of the design. Fortunately it is not necessary to unravel the threads in order to analyze the structure of the design. The course of each thread can be followed in detail. Likewise one may trace the movements of a behavior-pattern and follow the phases of its development from the origin to the consummation without isolating it from the conditions in which it thrives.

Any activity is composed of many ingredients, some of which may for convenience be considered as external and others as internal with respect to the organism, *but none of these factors can be considered as external to the behavior.* It is their relationship to each other which gives the activity pattern and form. A watch in the visual field is just as much an integral part of the activity of reaching and prehension as is the flexor-extensor movement of the arm. Although retaining their own identity, these factors unite in the formation of a behavioral pattern which is distinct from the summation of its parts. While the watch, the eyes, and the arms may remain as constant ingredients in the activity, as the behavior-pattern matures it becomes apparent that the interrelation has materially altered. The form and design of that interrelation have changed. These changes are the signs of development.

In order to ascertain these changes in a pattern of behavior, it is desirable to follow its course of development from the moment of origin until it attains functional maturity or fixity. Unfortunately when studying behavior-patterns in the human infant it is not possible always to realize this ideal. It is not possible because the human infant is a complex animal, born into the world with a long phylogenetic history plus a good start toward ontogenetic growth. Ordinarily the human infant is not available to the experimenter until after birth. All behavior-patterns do not, in ontogeny, begin developing at the same time. By the time the infant is born, some have not only started growing but have attained a high degree of fixity. It is therefore obviously impossible to start analysis of these action-patterns at their inception. Instead, one must calculate maturational level of the behavior pattern at the time it becomes available for observation.

The maturational level of each behavior-pattern should be calculated in terms of its phylogenetic history and of the neurostructural level at which ontogeny is controlled. Phyletic behavior-patterns have been considered here solely in terms of their ontogenetic manifestation. Since the cortex of the newborn infant is not functioning to any appreciable degree, all of the activities of the newborn infant are presumably controlled at an infracortical level. Some infracortical activities have not only attained fixity at the time of birth but continue to function at an infracortical level throughout ontogeny. Other behavior-patterns of the newborn infant may be in the late stage of development as infracortical activities and may soon appear in new forms as patterns of higher order. Such phylogenetic behavior-patterns are seen in the newborn baby merely as rudiments. These rudiments are controlled at an infracortical level and in the human subject they never attain functional maturity as infracortical activities. Although these rudimentary patterns never attain maturity at an infracortical level, they are succeeded by activities of functional similarity controlled at a higher structural level. Ontogenetic manifestation of still another order of phyletic behavior-pattern occurs in the human subject during post-natal development. It is possible therefore to study the ontogenetic manifestation of these patterns from inception, although they are rooted in the phylum.

While it is not possible to initiate study of all behavior-patterns of the infant at the time of their inception, it is possible to calculate the particular phase of development in which the action-pattern happens to be at the time the baby becomes available to the experimenter and, to continue study of its course of growth until it achieves fixity. The stage of development a particular pattern may have achieved can be estimated in terms of the degree of fixity it manifests and the degree of individual control over the performance.

Having used the growth of behavior-patterns as a means of studying the process of development, we have become convinced that the principles of growth are fundamental as to the process, regardless of the particulars of the growing organism. This analysis of growth in behavior-patterns is quite in line with observations which have frequently been made upon organic development.

Wherever there is development or growth there is also a period of inception, incubation, consummation, and decline. Nothing springs forth full-grown. Development is extremely gradual. There are spurts and rhythms in development, but it can safely be said that nothing is created without preparation.

> Growth is not a straight line affair nor is it a smooth curve, but it increases by steps, sometimes by leaps and bounds; at times it is slow and then again it is rapid. This is true not only for the body as a whole but also for its parts and organs, or functional systems. The cells themselves do not all divide and multiply at the same time, but first one and then another becomes active, and those that divide consecutively are not usually the adjacent ones (Bean 1924:45).

[Bean] was speaking of organic growth, but the statement could easily be used to describe the growth of a culture, an idea, or a behavior-pattern if the appropriate words were substituted. There are rhythms, spurts, fluctuations, pauses, and regressions preceding fixity, or functional maturity. The secret of a full understanding of the meaning and process of development lies hidden in the factors which determine these rhythms and fluctuations of growth.

A more detailed study of the development of particular action patterns than this investigation has afforded is essential if the nature of rhythmical fluctuations is to be ascertained. A behavior-course is not an isolated unit growing in every direction all at once. It is comprised of many aspects, each of which has its own growth rate and rhythm while each aspect of a growing action-pattern has its identifying way of developing, it is at the same time an integral part of the total behavior-pattern, in the same fashion that the behavior-pattern in question is an integral part of the total action-system of the individual. One aspect of a behavior-pattern goes through a period of rapid development, then pauses as another aspect moves rapidly forward. But the growth of each aspect of development influences and determines the growth of the other. The development of one aspect overlaps with the development of another so that there are no sharp lines of demarcation separating the phases of a developing pattern, but the connection of one phase or one pattern with another is more than a mere overlapping. There is a close interdependence in the growth of the various aspects of a pattern. Development works back and forward; here and there it strikes rapids, in other spots it pauses or regresses. The appearance of a new movement or aspect of a pattern facilitates or inhibits the growth of a previously developing movement and also determines the emergence and organization of a succeeding one. It is the gradual twining and interweaving of movements and phases of developing patterns which make it difficult to allocate the rhythms and spurts of growth.

In his discussion of an experience Dr. [John] Dewey has given an accurate description of the processes of development in behavior. It is natural this should be so, since any activity which he would call an experience must constitute for the experience development also. "An experience," he says,

> has pattern and structure because it is not just doing and undoing in alternation, but consists of them in relationship. ... In such experiences every successive part flows freely, without seam and without unfilled blanks, into what ensues. At the same time there is no sacrifice of the self-identity of the parts. ... As one part leads into another and as some part carries on what went before, each gains distinctness in itself. ... Because of continuous merging, there are no holes, mechanical junctions, and dead centers when we have an experience. There are pauses, places of rest, but they punctuate and define the quality of movement. They sum up what has been undergone and prevent its dissipation and idle evaporation. Continued acceleration is breathless and prevents parts from gaining distinction. ... Experience like breathing is a rhythm of intakings and outgivings. Their succession is punctuated and made a rhythm by the existence of intervals, periods in which one phase is ceasing and the other is inchoate and preparing (Dewey 1934:56).

While we have not in this investigation been able to ascertain the definition of rhythms and spurts of development, our findings, though tentative, suggest that the following phases or alternations can be expected in the growth of any behavior-pattern despite the complexity of the activity or the structural level at which it is controlled.

(1) Since we made observations only on overt manifestation of behavior, the initial phase in the development of a particular behavior-course would be that period just prior to the first emergence of a somatic movement indicating the appearance of the growing action-pattern. Whatever may be the nature of the growth prior to the overt manifestation of the action, it would have to be analyzed through some other method than direct observation. It is, for example, conceivable that a growth is in progress which will later make it possible for the baby to respond overtly to a bell held before his face, but the nature of the development cannot be observed in the actions of the newborn infant. This is true because there is a period following birth when his movements have no perceptible reference to objects within the visual field. The initial phase in the development of an activity is that period which occurs just prior to the first identifiable movements of the behavior-pattern.

(2) The second phase is indicated by the first somatic movement which can be recognized as a developing aspect of the behavior-course. This movement

is inchoate and ephemeral and is coupled with, if not obscured by, diffuse general activities. It can be observed only infrequently, and then it is not carried through to completion.

(3) Little by little this partial, incomplete movement can be seen to become more definite and expansive. In fact it often becomes so expansive as to appear excessive or exaggerated. There is a tendency to overwork a newly developing activity. This tendency to exaggeration is expressed both in time and in extent of the movement. As the child begins to get control over an aspect of a pattern, the activity itself becomes the incentive for repetition.

(4) In due process of growth, however, the exaggeration of this particular movement becomes checked or inhibited by the emergence of another movement. The conflict of the two is likely to evoke greater diffuse activity for a time, but gradually the second movement reaches the period of rapid development until it too becomes overemphasized, often to the extent of being excessive. Often the second aspect or movement develops so rapidly that it temporarily excludes the earlier one. Ultimately the excess activity is eliminated until the movement becomes restricted to its most specific and economical form—usually somewhere between the two extremes. We so often think of growth as an expansion or accumulation of something. In reality elimination and regression are as essential to development as accumulation and expansion.

(5) Once a pattern or an aspect of a pattern has attained a certain degree of fixity or definiteness it may unite or integrate with another aspect of the pattern or another action-pattern in order to form a new, more complex behavior-pattern. The process of development may therefore be quite different when the behavior-pattern is in different stages of maturation. At an early phase the process may be a matter of eliminating excess motion, growth progressing from a general diffuse state to one of greater specificity, whereas in the later stages of development the process may be primarily that of constructing patterns of greater complexity. Within a given individual both processes are going on at all times, since different action-patterns are in different stages of development.

In the development of a behavior-pattern the outstanding fluctuations or spurts appear to occur in connection with conspicuous changes in the growth cycle. That is, when one aspect of the action is declining there often occurs an increase in the activity before it finally fades out. This feature of development has become familiar in colloquial speech as the "second wind" or "end spurt." On the other hand a newly developing aspect of the pattern may herald its forthcoming. There sometimes appears sporadic evidence of a new aspect of the pattern before it definitely manifests itself as a newly developing part of the pattern. Often these sporadic occurrences are followed by long pauses or regression just before the movement appears as a characteristic aspect of the behavior-course. This feature of development has become proverbial in the familiar adages, "A new broom sweeps clean," "Beginner's luck," etc. When the same spurts and variations occur in the development of primitive behavior-patterns one becomes convinced that the principle is fundamental.

THE MODIFICATION OF INFANT BEHAVIOR

In fact and in principle the underlying processes which obtain in the development of behavior also apply to the modification of behavior. *Development* is a process of *modification,* and it is really impossible to draw a distinction between the two; but when we speak of modified behavior we usually mean that it has been modified in such a way as to change it from a course of development which occurs under the usual circumstances. It is in this sense that we have used the term *modified behavior.* Behavior-patterns can be modified by (1) speeding up the growth process, (2) prolonging or reducing the rate of development, or (3) in some way altering the form or sequence of the pattern which ordinarily occurs during development. Changes in any of these ways may be brought about through deliberate manipulation of the factors involved in the growth process. In so doing, it is not the principles of development which are altered but merely the relationship between the ingredients which make up the growth compound. There are many ways of artificially or deliberately altering the relationship of the factors which activate the growth process. One very obvious method is to increase or restrict the amount of exercise or use the growing behavior-pattern is allowed. In this way it is possible to estimate the effect repetition of function may have upon its development.

The extent to which exercise of an activity may alter the development of a particular behavior-course in infancy is contingent upon the following conditions: (1) the neuro-structural level at which the activity is controlled; (2) the state of plasticity or fixity of the behavior-course at the time increased exercise or use is introduced; (3) the state of fixity attained by the behavior pattern at the time the factor of special exercise is withdrawn, and (4) the phylogenetic origin and importance of the behavior-pattern.

Those behavior-patterns which have achieved a high degree of fixity and are controlled at an infracortical level are subject to no appreciable alteration through mere repetition of the activity during the post-natal development of the subject. Also phyletic rudiments of behavior-patterns controlled at an infracortical level are resistant to influence or alteration of any significance by increased exercise of the activity. Those phylogenetic activities which succeed these infracortical rudiments, that is, the kindred activities which are governed at a higher structural level, can be modified in minor details through individual exercise of the function. But the essential form and nature of the behavior-course is resistant to alteration through repetition of the function within the lifetime of a single individual. The degree to which the activities of this second order are modifiable in this manner appears to be in direct ratio to the temporal gap between the two types of activity. That is, the gap between reflex swimming movements and voluntary swimming movements is greater than is the corresponding relation between reflex stepping movements and voluntary walking. Therefore voluntary swimming is subject to greater modification through exercise of the action than is voluntary walking.

Activities of ontogenetic origin can be greatly accelerated through exercise of the performance, but the degree to which they can be modified is dependent upon the state of maturation or plasticity of the behavior-pattern at the time the factor of exercise is introduced. In general it appears that the period of greatest susceptibility occurs when the behavior-pattern is at its threshold of rapid development. In this connection it must be recalled that behavior-patterns do not grow all at once but that each aspect has its own period of rapid development. Therefore if maximum profit is to be gained by exercise of a growing behavior-pattern, the exercise should occur in connection with the particular aspect of the pattern which is in a state of rapid growth.

Since growth is gradual, first one aspect and then another gaining ascendancy, the exercise of the activity should be introduced in a similar way if optimum results are to be obtained. Since a child does not acquire all aspects of an action-pattern at once, it would seem reasonable that a gradual, step-by-step presentation of the activating agents should facilitate the development of new activities.

While there are critical periods in the development of any behavior-pattern when it is most susceptible to modification, it must not be inferred that behavior-patterns can be modified through exercise only during these critical periods of susceptibility. It means merely that these are the most economical periods of achievement. To delay beyond the period of greatest susceptibility means that other factors which have begun to grow will act as interferences and distractions, thereby rendering the achievement of the particular pattern more difficult.

The permanency of the expansion which an action-pattern gains through additional exercise is contingent upon the degree of fixity the behavior-pattern had achieved at the time the modifying agent, i.e., the factor of special exercise, was withdrawn. It does not necessarily follow that a performance which has been developed under special conditions will be retained after those conditions are removed. Unless the behavior-pattern has become fixed, it is only reasonable to expect that there will be a loss in performance unless the conditions which brought it about are continued.

Correspondingly, if the growth of a behavior-pattern has been hindered through restriction, it is to be expected that recovery will be evident when the restrictions are removed. In physiological development this aspect of growth is called regeneration and in behavior it is often referred to as reeducation. Great improvement in the development of behavior can be effected provided the cycles and rhythms functioning in the growth of particular behavior-courses can be disclosed, and provided the proper activating influences can be brought to bear upon the growth process at the most opportune time.

Not only in regard to the practical means and methods of modifying behavior development, but also in understanding theories of development themselves, we have often been confused because of the tendency to disregard the different degrees of plasticity or maturity of the growing organism in interpreting and formulating theories of growth. For example, the apparent controversy over individuation versus integration is clarified if the different stages of development of particular action-patterns are taken into account. From the descriptions of behavior development contained in the preceding chapters it is easily seen that at one stage of development the process of growth is predominantly a matter of eliminating waste motion, development being from an undifferentiated state to a more specific one, or a process of individuation. But once a pattern, or an aspect of a pattern, has attained an appropriate degree of specificity, further development is indicated by an integration of two or more action-patterns, or aspects of an action-pattern, into another of greater complexity. It is not therefore a question of one theory being correct and the other wrong. The two processes are by no means mutually exclusive. Actually, both processes can be observed in the actions of the same individual at the same time, but the processes represent different stages of maturation in the growth cycle.

This attempt of the writer to describe and interpret the course of behavior development is an excellent illustration of the stage of incubation represented by an inordinate amount of excess motion, a tendency to overemphasize some particular aspect just because it is a slightly new point of view to the author. Somewhere in the diffusion there is recognized a deliberate and conscious movement which, though inchoate, is nevertheless in the direction of a goal. In a descriptive analysis of development one is beset by difficulties, among them the most obvious being the fact that development is *movement,* dynamic movement. The static bounds of language concepts are too limiting to allow a precise definition of development. Perhaps the time will come when the movements of growth can be expressed in mathematical formulas as precisely as the movements of celestial bodies, but until that time arrives we shall have to be content with cumbersome descriptive analyses.

✂ 11 ✄

Individual Development

The data obtained in these studies of the development of behavior have been examined by Campbell and Weech (1941) in order to determine whether or not they contain criteria which are capable of characterizing the individual child against a background of his peers. The techniques for manipulating the original data in order to determine these criteria, or measures of development, are fairly complicated. Since at this time we are interested in the conclusions to be drawn from their results, the actual techniques need not concern us. It is at once apparent that the method pursued throughout the studies, being one which has dealt with repeated serial observations on the same child, should be capable of yielding information concerning the trend of development which is appreciably more accurate than could be secured by occasional observations at a few stated intervals along the age axis. At the same time the intense concentration on the individual child has precluded the recording of data on a large number of activities. It will be clear that the failure to include such a large group of different behaviors for each child introduces an element of uncertainty which would affect any serious attempt to set up for each of these children a measure of development in neuromotor activity of the same type as is provided by the conventional intelligence quotient. The study, however, is of some importance, not only because it demonstrates the feasibility of establishing such a developmental scale under more appropriate circumstances but also because the nature of the results permits an interpretation which is in harmony with the concept which relates the development of behavior to the growth of the cerebral cortex.

The data selected by Campbell and Weech for their study consisted of observations on forty children, including five sets of unisexual twins. The observations began during early postnatal life and continued during the sequential developmental phases leading to mature behavior in the activities of

Reprinted from Myrtle B. McGraw, *The Neuromuscular Maturation of the Human Infant*, Copyright © 1943, Columbia University Press, New York. Reprinted with Permission of Columbia University Press, New York. Reprinted by permission of Mitzi Wertheim.

sitting, creeping, walking, reaching for a lure, and reacting to the prick of a pin. The analysis was made by two methods which utilized wholly independent properties of the data. The fact that both methods led to identical results in describing the emergence of traits characterizing each individual child as age advanced, enhanced considerably the significance of the results.

Without making use of mathematical representations, the principles involved in the analyses can be presented in a fairly simple way, in a way which permits visualization of the criteria of development used in this study. Let us begin by considering a single behavior skill such as creeping. Forty children were studied at repeated intervals throughout at least the first six or seven months of life. The achievement, or amount of development, at any age for each child can be expressed in terms of the age at which he attained a specified degree of development. Let us think of the situation at the time of birth by drawing an analogy to an upright ladder made up of forty rungs. Each rung of the ladder represents the position of an individual child as in relation to the thirty-nine other rungs, or children. The top rung of the ladder represents the most advanced child in the series, that is, the child in whom the most advanced stage of the activity was observed at the earliest age. Similarly, the bottom rung of the ladder refers to the least advanced child in the group, that is, to the child who was older than all the other children before he reached the same criterion of development. The remaining rungs of the ladder describe the position of the intervening children. The actual techniques involved in determining the relative positions of the various rungs are fully described in the paper by Campbell and Weech (1941).

Now let us imagine a series of similar ladders set up at monthly intervals throughout the age of infancy. The development of the individual child against the background of his peers can be followed in terms of the position of his rung in all the ladders. It is clear that the changing events which attend development will lead to shifts in the rank order distribution of the ladder rungs as age advances. Such shifts were actually observed. It was felt at the outset that the shifting in position would probably not happen in a haphazard manner. Actually, it was demonstrated that the position of any one rung in all the ladders could be described by a straight line when the ladders were placed at appropriate intervals along the age axis. Which is to say, that a child who was developing with unusual rapidity during the early months of life, would continue to exhibit approximately the same velocity of development throughout the period of study.

Consideration of these ladders permits the isolation of several components of development. We may think first of the position of a child's rung at the time of birth, and this may be looked upon either as a measure of hereditary endowment or as indicating the extent to which maturation had progressed during life *in utero*. Next, we can set up a criterion of development in terms of the slope of the line which describes the position of each ladder rung in the series of ladders. This criterion can be looked upon as a measure of the velocity of development. Finally, we can set up a criterion for each age based simply upon the position of the child's rung at that age. This last criterion is a measure of achievement and depends both upon the child's position at the time of birth (criterion 1) and upon the velocity of change (criterion 2) up to the designated age.

For simplicity the considerations in the foregoing discussion have been concerned with a single behavior activity, namely, creeping. It is clear that analogous criteria of development can be set up for each of the activities included in the study. With these data at hand, it became possible to ascertain the extent to which each one of the criteria was capable of identifying the individual child. In general, one would expect that a child who learned to creep sooner than the average would also exceed the average in learning to walk, to sit, and so forth. The success of any one criterion in identifying the individual child can be assessed in terms of the consistency exhibited by this criterion throughout the entire series of behavior activities. For instance, a child who in learning to creep occupied the position of rung 2 in the first ladder of this series and the position of rung 8 in the six-month ladder, and who occupied similar positions, respectively, in the ladders representing the other behavior activities would be accurately identified as an individual with regard to his development. However, if the positions occupied by any one child were wholly and randomly different for each activity, it is clear that the criteria would be yielding nothing descriptive of individual characteristics.

With the above principles of study in mind, the results can be stated briefly. It was found that the individual child could be identified, but not accurately identified, by the state of his development at the time of birth. Individuality is more clearly expressed by the criterion which depends upon the velocity of change with advancing age. The best and most significant indications of individuality were yielded by the criteria which described achievement at progressive age periods. It has been said that individuality was poorly defined at the time of birth. By the age of fifty days, however, the analysis of the criteria of achievement indicated a somewhat higher degree of characterization of the individual than was secured from the study of velocity of change. With each advancing age period up to the age of one hundred and fifty days, the evidence clearly indicated the emergence of an increasingly accurate characterization of the individual. The nature of the data suggests that the manifestation of individuality probably continued beyond the age of one hundred and fifty days, but because less and less data were available as the children grew older, it was not wise to extend the conclusions beyond this period.

In the second method of analysis a wholly different property of the data was utilized; namely, advantage was taken of the circumstance that the forty children included five sets of unisexual twins. Again we can think of the method in terms of the position of ladder rungs in a series of ladders. In this case, however, the position of the rungs of each ladder at different ages was determined from the average performance of the child in the entire series of behavior activities. A study was then made of the position of the ladder rungs corresponding to the unisexual twins throughout the age series of ladders. Characterization of the individual in terms of achievement could be considered to have reached a maximum value if at some stage of development each pair of twins occupied adjacent positions on the rungs of the ladder at a given age. By the use of contingency tables and the calculation of chi-square from them it was possible to obtain quantitative measures of the significance of adjacent positions of the twins at different age periods. This method of analysis likewise revealed an emergence of individuality which progressed with advancing age up to the age of one hundred and fifty days. By this time a rank order arrangement of the forty children disclosed that twins were in adjacent positions in four out of five instances; in the fifth instance the twins were separated by only two other children.

At this point it appears worthwhile to comment on the two methods of analysis used in this study. By the first method the variability of each child throughout several behavior activities was assessed and given meaning in terms of characterization of that child according to his variability throughout the entire group of children. In the second method measures for each child in the several activities were averaged and the average values were arranged in rank order of magnitude; the faithfulness of the measures in characterizing the individual was then determined by the relative position of twins in the rank-order scale. It is true that both methods yielded the same conclusion regarding the emergence of traits characterizing the individual as he advanced in age; however, it is worthy of note that the significance of these findings as expressed in terms of conventional statistics was far higher by the second method, which took advantage of the inclusion of twins in the study, than by the first. This fact should be noted by all investigators of the phenomena of development. The original expenditure of time required to locate twins who can be included in a study will be repaid many times by more significant results, or if only an equal degree of significance is obtained, at least there is a tremendous saving in the total expenditure of time and labor.

With regard to the relation between overt behavior and development within the nervous system, the implications of the present analysis are clear. Reflex behavior during the early postnatal weeks is controlled chiefly by subcortical centers in the brain. The centers which relate to the reflexes involved in early sitting, creeping, and walking behavior are different and relatively independent. The small amount of association within individual infants which could be demonstrated can be attributed either to parallel development of different centers under identical environmental influence, to the functional overlapping of the centers involved in the several behaviors, or to an outside influence which exerts control over all the centers. If the last possibility is true, it may be that even at this stage the cortical centers are beginning to function. In any case, as the infant grows older and as maturation in his cortex progresses the influence which coordinates behavior becomes stronger. The circumstances allow one to think that the cortical centers themselves influence this change.

❧ 12 ❧

Let Babies Be Our Teachers

One of the most significant recent invasions of medicine is the well child. Particularly in pediatrics the doctor is no longer content to focus attention solely upon the disease germ which has invaded the body. A large part of modern pediatric practice is simple pedagogy. A mother seeks a doctor's advice concerning the child's behavior and misbehavior as well as when he has an elevation of temperature. It is reasonable, therefore, that the pediatrician should turn to other fields of endeavor—to education, psychology, and psychiatry—for answers to some of the questions with which he is daily confronted. So-called "normal behavior" has never, and still does not, occupy a large place in the medical school curriculum. To compensate for the shortcomings of our professional training the doctor, the educator, and the psychologist must learn to understand each other's thought-ways. Theories evolved in any one of these fields find expression in the practice of another.

Pavlov evoked the secretion of the dog's salivary glands to the ringing of a bell. Watson filled a human infant with terror at the sight of a furry toy rabbit merely by associating it with a loud sound. Thereupon the "theory of conditioning" became popular and in many ways invaded the daily habits of our homes. Pediatricians began advising young mothers to start "conditioning" their infants in toilet habits at the tender age of three weeks, or as soon as they were home from the maternity wards. Theoretically, the bringing up of children looked simple under the conditioning theory. All one had to do was condition the child positively to those things desirable and negatively to those that were not. The only trouble with it was that life is not so simple as that; in practice the theory just didn't work.

Earlier, at about the turn of the century; [Edward] Thorndike, then at Harvard, had begun to study the way in which domestic animals solved certain problem-box situations. He discovered they did so by "trial and error."

He formulated the now famous laws of learning—"The Laws of Use, Disuse, and Effect." They immediately became stock material in educational and psychological textbooks. For years thousands of teachers journeyed to the Mecca of Teachers College to learn how to put these laws into effect in the classroom. These illustrations are not mentioned with any thought of casting doubt on the merits of the experiments or the soundness of the theories derived from them. They are mentioned merely to emphasize that theories based upon laboratory experiments cannot be tossed off as just so many intellectual gymnastics. They do influence our ways of life. The white rat, running mazes, and doing other trick performances, has provided material for many theories and amplified many others which have influenced our thoughts about, and our treatment of man. While we may sometimes want to call a man a rat, no one would, for a moment, pretend that a rat is a man.

But it is not necessary here to point out the dangers of adapting a theory, evolved in the study of one species, to the treatment of another. These dangers are well recognized and are becoming more so. Unfortunately, however, the dangers of applying a theory derived from studies of a mature animal to an immature one, regardless of the species, are not equally well recognized. Few experimenters bother to report the maturity of the rats upon which they experiment and from which their theories are obtained. Yet the maturity of the experimental subject is most important, especially in investigations leading to theories of behavior. A mature rat behaves differently from a young one even if you merely pinch his tail. It is conceivable that a similar differences may prevail in their manner of running mazes or discriminating between a square and a triangle.

It has always seemed to me unfortunate that so many of our educational theories have been derived from studies of animals or humans whose nervous systems were mature at the time of the original observations or experiments. Not only from the experimental laboratories but from clinical practice, too, many of our interpretations or theories of child nature have been based upon adult performances, or even adult abnormalities. Freud's generalization, for example, that the symptoms of the oral pervert may be due to the repression of "perverse" fantasies has led to some of the most fatuous assertions concerning the normal sucking habits of infants.

We who have lived through the medley of concepts about child nature during the past twenty-five or thirty years, and the different methods of training and management employed, can now take heart as we see a budding amalgamation of many different schools of thought. I, for one derived great satisfaction from a remark of Dr. Ives Hendrick before a meeting of the New York Psychoanalytic Society, when he said,

> I shall propose the thesis that psychoanalysis has neglected the overwhelming evidence that the need to learn how to do things manifested in the infant's prac-

tice of its sensory, motor, and intellectual means for mastering its environment is at least as important as pleasure-seeking mechanisms in determining its behavior and development during the first two years of life.

When I think back on the various concepts of child nature and management, I like to imagine how different our textbooks on the subject might have been if men like the late Dr. George Coghill [Coghill died in 1941] had been intent on selling their wares as well as on being pure scientists, or even if they had been sought out by the educators and practitioners and had not been allowed to isolate themselves in their own research laboratories. Coghill set for himself the task of determining the correlation between the development of patterns of behavior with the progressive differentiation of the organs, especially the nervous system, which execute the behavior. For forty long years he sat in his laboratory, unheralded and unsung, patiently pursuing his goal as it is demonstrated in the lowly salamander, the Amblystoma. As he observed the changing behavior of the salamander from early embryonic life to adulthood, and as he ascertained the changes in neural structures which accounted for these changes in behavior, he became convinced that there are two fundamental processes of organic growth. These two processes involve the question of relationship between the total organism and its members, or the parts of which it is composed. They are best revealed in his description of early embryonic life. In the salamander embryo there is an initial period when the leg, even the forelimb, can move only if and when the spinal axis bends. The motility of the appendage is completely subjugated to the movement of the total body, and is incapable of independent movement of its own. During this period of development if you touch the limb of the embryo with a stiff bristle you get a response of the total organism, a bending of the spine to one side. Later the leg becomes what Coghill calls "individuated," that is, it becomes capable of independent action. Then the young salamander is able to move the leg forward or backward without necessarily bending his head and spinal column. But he can enjoy that freedom of leg action only if the leg remains a part of the body, and if its movements do not interfere with but are for the benefit of the total organism. Coghill has also shown that the development of the independent limb actions is antagonistic to the primary one, because the activity of individual appendages tends to break up the whole movement into parts and to endow these limb or part movements with an individuality of their own. Antagonism between two different processes appears to be a fundamental property of growth.

In one of his later and more theoretical papers, Coghill attempted to show *how this antagonism between the primary or fundamental patterns of behavior and the development of local or individual action may constitute the organic basis for psychic conflict.* It is not within my province to explain the anatomical basis for this application of his theory, and it is a little difficult for us to think of a salamander being guilty of psychic conflict, suffice it to say that it involves the relationship between the total organism and the freedom of its parts. In normal development the total organism should retain sovereignty over the partial or local patterns of behavior. But, if the total organism fails to allow freedom of action to the members, normal development cannot be achieved. On the other hand, if the action of individual members is such that it does not contribute to the welfare of the total organism, then behavior becomes abnormal.

In several of his papers Coghill attempted to explain reflexes, instincts, and the conditioned reflex in terms of his theory of primary integration and the individuation of partial patterns. Yet never, so far as I know, did he draw a parallel between early embryonic development and the evolution of social orders. I would like to digress for a moment—infringing upon his rights—to point out that the process of growth of social systems, social organisms, is not altogether unlike that of biologic organisms, as exemplified in salamanders. In doing so I hope I can impress more clearly the theme of this lecture, namely, that *we should turn to biologic growth processes for suggestions and theories about the way to mold the behavior of individuals and of groups.*

A few years ago I was reading a most entertaining and stimulating little book called *The Soul of the White Ant* by Eugene Marais (1937). He undertook to show how the organization of the termitary, the house of the ant, was similar to biological organization, the organization of our own human bodies. The function of each individual ant was highly specific. Some served as guards, others as white or red corpuscles, each moving about with individual freedom but always for the benefit of the total structure, the termitary. One day, when discussing the book with one of the doctors of Presbyterian Hospital, I dwelt upon the parallel between biologic and social development with perhaps more than justified enthusiasm. To my amusement he remarked, "Yes, and Hitler thinks he's the pituitary gland." At that time Hitler—sheltered in his *sella turcica* at Berchtesgarden—was, through the Munich Pact and other moves, controlling the affairs of man.

Seriously, speaking of Hitler and the Nazi regime, can't you see a parallel between that and the behavior of Coghill's salamander embryos? The individual cannot move or act except as the spinal axis bends. If you talk to a Nazi, a real Nazi, you cannot get a response except in terms of his identification with the total system. Freedom of action of members in that social system has not been attained; to use Coghill's terms, they have not become "individuated." We are too close to it, as yet, to know whether these are merely the signs of embryonic development, since as a social organism Germany is still young, or whether they reflect abnormal growth—the failure of the primary mechanism to allow some freedom of action on the part of its members.

In contrast, we in America are like the salamander who has just reached that stage of progression which permits freedom of action of the parts. As individuals, we can move independently. But like the young animals at the onset of locomotion, our movements are ataxic, disjointed, and incoordinate, especially in so far as our concept of the relationship between the individual and the social system is concerned. We haven't learned to make our right arm and left leg move synchronously together while an eye keeps steadily on the goal in the march forward. We haven't learned to make our individual freedom serve most appropriately for the welfare of the social system; the social organism.

The parallel between biologic or organic growth processes and social and educational development can be even more convincingly observed in the early behavior development of our own human babies. This is true because the baby comes to us at the end of a long phyletic history and also because at the time of birth his nervous system, particularly the brain, is immature. Therefore, during the early years of life we have a chance to watch how the brain develops and to take our cues for education therefrom.

It was some twelve years ago, under the supervision of the late Dr. Frederick Tilney, that I embarked on a study of the behavior development of the human infant. He hoped, at that time, to accomplish the sort of thing in our studies of the human which Coghill has so admirably done in his investigations of the salamander. We know now that, because the human is so much more complex, it will require more than the professional life-span of one generation to achieve that objective. While we have fallen short of the goal, some of the by-products of that undertaking have been both gratifying and entertaining. During the early days of 1931 it was difficult for me to make the pediatricians of Babies Hospital, with the singular exception of your farsighted Director, Dr. Herbert Wilcox, think that the ringing of bells and blowing of whistles before newborn babies could be serious business. Now I can tell you that most of the doctors of that same hospital—perhaps not because of our antics but certainly not despite them—look upon the wriggling mass of a healthy newborn with respect, respect for the things he can teach them—not merely with an eye as to what they the parents can make out of him.

Although after a decade of intensive study we are still unable to identify specific neural structures which account for particular patterns of behavior, the gradual unfolding of behavior development as exemplified in the motor development of the infant has taught us a great deal about the process of growth. In the meantime enough progress has been made in the investigations of cellular structures of the fetal and infant brain to make the interpretations of our behavior data more meaningful. In this connection we have leaned heavily upon the contributions of Dr. J. LeRoy Conel (1939a; 1939b) of Harvard.

As a scientist, Dr. Conel is another Coghill. He sits patiently in a corner of his own laboratory, meticulously working out the details of the anatomy of tiny brain cells without giving a thought as to whether his work is recognized or ever touches the life of his fellow men. If we weren't busy with a war perhaps some brilliant young scientist, with a flare for the practical, could set himself the task of interpreting the details of Conel's findings for the benefit of those more actively engaged in molding the individual and the social system in which we survive. But that task will probably have to wait now for another generation. In the meantime if I, having drawn upon his work only to the extent of limited abilities, can demonstrate to you the advantage of using immature nervous tissue as our basis for the formulation of pedagogical techniques and an understanding of social growth, then I shall have achieved the objective of this study.

If at times, such as now, we find the complexities of social mechanisms incomprehensible and discouraging, we can take heart when we think of the complexities of our own brains which created them. Dr. Judson Herrick has put it graphically when he said, "The human brain is the most complicated structural apparatus known to science. If all the equipment of the telegraph, telephone, and radio of the North American continent could be squeezed into a half gallon cup, it would be less intricate than the three pints of brain that fill your skull and mine." Is it then presumptuous for us to think that the maturation of the most complex of all structures could shed some light upon the problems of mankind?

We have said that the baby provides the most promising material for the study of immature brain tissue. But a baby's skull is not an egg. We can't make a window in it cover it with cellophane, and watch the embryonic cells mature. While we can't do that, we can watch the changing behavior of the infant and know that those changes in behavior mean that the cells of the brain are ripening and getting ready for action. Furthermore, studies of the cellular changes as determined by histological analysis have already progressed sufficiently to provide a framework for the interpretation of behavior development.

For the sake of simplicity let us say that the human brain is composed of two major divisions—the cerebral cortex, and the subcortical nuclei. The cortex is that mass of grey matter which spreads over the two cerebral hemispheres. It is the cortex which most clearly differentiates the human from lower animals. Someone has said that "the most ungifted normal man has twice as much of this master tissue as the most highly educated chimpanzee." It is because this master brain tissue, this human cortex, is not functioning appreciably at the time of birth that the growing infant provides such rich material for the study of growth and development. It is also because the human infant occupies a peculiar position in the evolutionary scale that his development can reveal so much about the process of organic growth. He is in the position of not being able to count upon the old part of his brain, the subcortical nuclei, to serve him as effectively as it does the young domestic animal. At the same time the new brain, the cortex, is still too immature to be of much help to him. A young foal can, within a short time after birth, stand on his four spindly legs and walk with considerable certainty. This he does with an old, primitive brain mechanism. In so far as walking goes, the baby's primitive brain is not as efficient as that of the colt. Still the old brain is there, and there is enough activity in it for us to see in the behavior of the baby a residual of primitive walking. If you hold a newborn under the arms so he can just rest his weight on his feet he will often make stepping movements, simulating walking. But before he can really walk, these stepping movements become suppressed or dissipated for they are the property of the primitive brain.

In fact *it can be safely assumed that all the behavior of the newborn infant is controlled by the primitive part of the brain,* by the subcortical nuclei. Both Tilney and Conel are of the opinion that the new brain, the cerebral cortex, is not functioning to any appreciable degree at the time of birth. The cells of the new brain are all there, but they are in such an immature state they do not take an active part in the control of behavior. This information provides the basic hypothesis for the interpretation and understanding of later development of the infant. It means that when the newborn holds on to a stick suspending his body in midair, when he makes crawling and stepping movements, and even when he makes fairly coordinate swimming movements he does so with the primitive brain centers. It means that when new types of behavior burst forth, the cells of the cortex, the higher brain, are beginning to participate and take control of the child's activity.

The quality of behavior controlled by the cortex is different from that under the dominance of the subcortical centers. In cortical behavior the child exhibits some awareness of what he is doing, even if it is merely lifting his foot; cortical behavior is purposive and adaptable, less stereotyped than the responses of the subcortical centers.

One of the first things to strike the observer of infant development is the struggle that goes on between the two major centers of the brain as soon as the cortex matures enough to participate in any particular activity. Before the cortex, or the higher brain centers, can assume full power or control of any function it must suppress or curtail the activity of the lower brain centers. So in the infant one sees a disorganization of the primitive pattern of behavior, followed by a period of relative inactivity, and then the gradual expansion of deliberate, cortical functioning until some degree of integrated action is attained.

This struggle between higher and lower brain centers for the control of behavior is evident in all the early motor achievements of the infant. But it is
most readily demonstrable in the development of creeping because this is an
activity which ultimately calls for the synchronous working together of arms
and legs. The cortex does not take over the movement of all muscle groups
involved in this activity at the same time. It begins to participate in the movements of the shoulders and arms first, and then the legs. So for a time, at the
onset of cortical functioning, the child may pull and tug with his arms, trying
to move himself forward, while the legs hang along like a dead weight and
just won't play ball. When they do begin to participate it's only a half-
hearted affair. One leg goes this way and one that; an arm swings out twice
before a leg can be persuaded to move; and sometimes the child loses his customary control of shoulders and arms and falls on his nose. But, ultimately,
the body and all four limbs begin to work together with the precision of an
electric clock, and the child goes scooting here and yon without giving a
thought to the business of creeping. The act of creeping has reached maturity
and is therefore enlisted into service for the benefit of the total organism, for
the benefit of the baby.

And when I watch the way in which babies achieve a smooth frictionless
method of creeping and an easy, integrated gait, I'm inclined to feel that my
colleagues, the psychologists and educators, got off on another unfortunate
start in their early experimental methods and theories. They not only neglected to give adequate consideration to the maturity of the nervous systems
they were experimenting with, but the very nature of their experiments led
them to emphasize end-results, the achievements of their subjects, whether
they happened to be animals or humans. Their experimental procedures
were such that they were forced to appraise performance in terms of errors
made or time consumed in the subject's solution of a problem, the goal of
which had been determined by the experimenter. If I were a cultural anthropologist I would probably go on to say that they adopted these particular
techniques because they fitted in with and reflected our American ways of
thinking. We are a success loving people. We love success so much we are often inclined to forgive the methods by which it is achieved.

But I'm not a cultural anthropologist and, therefore, am content merely to
say that in the development of infants it is not so much the achievement of a
goal as the mode of performance which most clearly reflects advancement.
In the development of creeping for instance, the infant who makes no progress whatsoever along the floor, the one who never retrieves the alluring lol-
lipop, but gets on his hands and knees and rolls back and forth, may be further along developmentally than the one who pulls and tugs with his arms,
dragging his body behind until he finally reaches and devours the tempting
lure which was only a few feet away. The one who gets on hands and knees,
though he never reaches the goal, reflects a more advanced stage of develop-

ment because his behavior indicates that the cells of the motor cortex which control the movement of the legs are trying to cooperate with those which control the movement of the arms.

Another impressive lesson the serious observer will learn from the baby is that progress and development do not move in a straight line. An achievement gained does not necessarily mean an ability retained. There are losses and regressions in performances. Of course, every mother dealing with the practical affairs of bringing up children knows this. But because it has not been emphasized in our theoretical concepts of development and learning, because we tend to think in terms of achievement and success, these periods of regression are not always accepted with equanimity. This is especially so when the performance is one which we desire to cultivate in our children. How often does a pediatrician hear a mother complain that her child was "beautifully trained," but that he has backslidden, and she's worried lest the fault is hers, her poor management. If under such circumstances both the mother and the pediatrician could recognize that regressions are a natural process of growth they would not aggravate this phase by anxiety. Instead they would look for the budding signs of maturation of some new ability, and bide their time. It may be that those cells of the child's cortex which give vent to his vivid imagination have just begun to ripen and he can't make them dovetail with the maturer cells which had earlier enabled him to respond to physiological urges. This type of incoordination between the development of two seemingly unrelated functions is not dissimilar to the creeping baby who can't quite make his right arm and left leg move synchronously in a well-integrated creeping pattern. Within the course of events this period of incoordination between the development of two functions is overcome, especially if it is not distorted by tensions and anxieties.

The obligation of future genetic psychologists, it seems to me, is to ascertain the signs, the "behavior syndromes," which indicate that a child's maturing nervous system is ready for a particular type of activity. These signs are not to to be found exclusively in chronological or mental age ratings. We all know that one infant may begin to walk at nine months and another at eighteen months, and, when they are three years old, neither you nor I can tell the late walker from the early one.

The signs of neural readiness are to be found in the child's mode or manner of behavior. Already a pediatrician need no longer tell a mother to start training her infant in toilet habits at the tender age of three weeks, in keeping with the "conditioning theory," nor at ten months or fourteen months. He can tell her to start training when the baby begins to exhibit an awareness of what he is doing, when his behavior indicates that the higher brain centers are beginning to participate in the function.

Parenthetically, may I say that if in the course of this discussion I have drawn literally upon training in toilet habits I have done so for two reasons.

One is, it happens to be a function to which we, at the Babies Hospital, devoted considerable attention, and the other is because it always remains, even in adulthood, a kind of partnership between the cortex and the subcortical nuclei, in so far as the control of performance goes. There are lots of partnerships of that order which are exhibited in our little everyday habits. A newborn baby can cough, sneeze, and yawn about as efficiently and in essentially the same manner as you and I. In so doing we all exercise the lower centers of our brains. There are many times when in doing these things, even as adults, they are still fundamentally under the control of subcortical centers. A bit of saliva, a drop of water, or a crumb on the epiglottis may set me to coughing, and I couldn't help it even if it proved embarrassing at this particular moment. But our Chairman, Dr. McIntosh [Director of Babies Hospital, Columbia University, 1931–1961], may surreptitiously take a look at his watch, cough, and give me that well-known glance. Then we would all know that the time is getting late, and I should hasten toward an end. In that case the cortex would participate in stimulating the act of coughing, though the motor mechanisms involved would be essentially the same as those of reflex action.

The reason why we should ever keep our eyes open to detect when the cortex is beginning to participate in a particular function is because this seems to be the time when the child can most economically improve or profit by training or education in that activity. In saying this, I turn again to the early organic behavior development of the infant. He provides the cues.

When an infant is just becoming able to do some act he shows an indomitable urge to exercise it. The baby who has just learned to turn over can hardly be kept on his back; the one who is able to pull himself up by the bars of his crib will do so repeatedly, even though he is in misery once he gets there because he hasn't at the same time acquired the ability to let himself down; and the baby who is expanding his sensory experience of objects makes a nuisance of himself by emptying the contents of every bureau and cupboard in the household. The obvious lesson for us in this drive of the infant is to provide opportunity for exercise and nourishment of each function just as it is beginning to emerge as a part of the child's conscious behavior. If proper nourishment and challenge can be provided at these critical times, a child's performance in a given activity may be extended beyond that normally expected. An infant feeling the urge to exercise the neural mechanisms of creeping will enjoy the challenge of climbing inclined planes, and in short order will be scaling those of surprising steepness. One just learning to walk independently will also learn to roller-skate, because roller-skating adds challenge and zest to the maturing nerve centers which enable him to walk.

But alas! This urge to pursue and practice a given activity does not continue until it has attained ultimate proficiency. That is where we as parents, doctors, educators, and psychologists, with our eye ever on end-results, on achievements, are at odds with the natural course of development. Once a child has achieved even a moderate degree of efficiency in one performance, his interest shifts to a new one. It is just as important to know when not to drive him to achieve in a given direction as it is to know when to stimulate and encourage the expansion of some performance.

In the course of development of any neuromotor function there are periods of struggle, conflict, and confusion; periods of drive and rapid advancement; of regression or inanition. But ultimately integration of performance is achieved. And once it is achieved, it acts not only for itself, but in the service of a larger constellation of functions. A child learning to read or write will show the same struggle, the same focus of attention, the same incoordination and drive as does the infant in learning to creep. But, once he is barely able to write, his focus of attention shifts from the act to what he wants to say. Improvement in penmanship nevertheless continues, ultimately becomes stereotyped. At last his focus of interest in what he says is shared with an equal interest in rhetoric composition. This gradual weaving of one pattern of behavior together with and in the service of another is clearly evident in the early organic development of the infant. There is give and take; first here, and then there; but in the end a working balance is attained if and when performance is normal.

It has been said earlier that those who are molding the lives of individuals and those who are concerned with the affairs of man would do well to look to the fundamentals of biology for guidance. It seems that there are fundamental properties of growth, and that the same principles are at work whether it is the growth of a salamander embryo, the behavior of an infant, or a social system. One of the outstanding properties of growth is struggle and conflict. Coghill saw it in his salamanders; we've seen it in the neuromotor development of the infant. It is fundamental; it is organic. We can expect it in the growth of social orders.

While conflict is indigenous to development, it should not be inferred that conflict in social growth must always take the form of military wars. Accepting the parallel between biologic and social growth, we can take a more hopeful point of view—perhaps not for our own generation but for that of our children and grandchildren. The reluctance to enter this war, the inertia to get it moving once it was formally declared—that is, the "sit-down war"—are signs that the idea of the abolition of war is beginning to penetrate our social consciousness, our social cortex, if you will. To me it is analogous to the period of relative inactivity in the baby's motor development which represents the period of transition from subcortical to cortical control of behavior.

Even now, when we are in the thick of battle, developmentally, it is not the antagonism between the Axis and the Allies that is most significant. These are merely symbolic, or perhaps the mechanisms of expression. The significant thing, developmentally speaking, is that we can't quite get rid of the old, primitive idea that we must fight wars over geographical boundaries, even now when geographical boundaries no longer mean anything. This old hereditary pattern of social behavior is in conflict with our budding consciousness of the relationship between the individual and the group, between one nation and the world. That the social organism is a global affair is just beginning to penetrate our social cortex. Once the social cortex takes command new patterns of behavior will emerge, new social systems will arise.

And when we remember Coghill's salamanders and my babies we know that these new systems must be deliberate, planned, and purposive; they must reflect the activity of a social consciousness, a social cortex. There are many indications today that this social consciousness is emerging. Consider, for example, our changing attitudes toward community health, or the meaning back of our rationing system.

If one adopts the biologic point of view in thinking of the present crisis one is confronted with two alternatives. In the process of evolution man has specialized on the development of the cerebral cortex. The dinosaur specialized on mass. His own specialization became his destruction. It could be, and I have often heard it argued, that the product of man's specialization, the use of the cerebral cortex for the invention of lethal instruments, will ultimately spell his total extinction. On the other hand, considering more dynamic biologic phenomena, one sees the periods of transitions and regressions, and takes hope. For me this is not only the more satisfying but the more convincing point of view, for the essence of cortical activity is its adaptability. I can't tell but feel that out of this confusion there will arise a purposive, planned, and finally integrated method of social living, both structurally and functionally.

If we follow the biologic analogy further we know that this must be a social system which recognizes the sovereignty of the total organism—and the total organism is now the world—but at the same time grants freedom of action to the appendages. We also know that the independent action of the various appendages must achieve integration, that they must learn to work together smoothly and purposively not merely for themselves but for the benefit of all.

The things said about the growth and development of infants are said with confidence, with as much confidence as any investigator can feel who does not deal with concrete material which can be measured in known units. The comparison drawn between organic and social growth is admittedly speculative. For me, at least, it reflects a broader horizon which I gained from the observation of babies, a horizon which increased my respect for the principles of physiological growth.

But even in my enthusiasm for those things biologic, I still can't say to you, "Go to the ant, thou sluggard, consider her ways to be wise." After all, her ways are reflexive, stereotyped. But I do say, "Look to the baby, you parent, doctor, and teacher: Consider his ways and learn."

PART FOUR

Psychobiology and the Interdisciplinary Study of Infant Development

Introduction: The Developmental Psychobiology of Myrtle McGraw

DONALD A. DEWSBURY

The four essays included in this section have in common a focus in developmental psychobiology. They reveal the sophisticated nature of Myrtle McGraw's thinking in this area and illustrate some of the recurring themes in her work as a whole. I will first provide a brief overview of these four essays and then draw out some of the themes that run through them.

THE FOUR ESSAYS

The first essay, *Basic Concepts and Procedures in the Study of Behavior Development,* is centered around McGraw's long-standing concerns with the methods used in developmental research. She addresses three sets of issues: the importance of concepts, the nature of organisms, and the longitudinal method. In each case, McGraw strives to delineate appropriate methodology but refuses to proscribe specific approaches. Development is complex and what may apply well in one situation may not work well in another. She therefore defines only broad guidelines; the specifics must be filled in and adapted as appropriate to the situation at hand. She begins by opposing the Baconian view that one can productively collect information in the absence of guiding concepts. Concepts, however, are viewed dynamically, changing with time and amenable to refinement and improvement. The influence of American pragmatism is apparent. Hers is a very contemporary view of the contextualization of the metaphors and concepts used in science. Even the definition of an "organism" will change with context. The "organism" may be bounded by epidermis, but it may be so small as a cell or so large as the universe. Which definition works best depends on the scope of the investigation. The theme is carried through with her treatment of the longitudinal method. There must be repeated observations of the same characteristic in the same individual, but the number and timing of these observations are free to vary so long as they "have meaning in the particular problem under investigation" (McGraw 1940:86).

The second essay is McGraw's (1946) classic review of literature on the maturation of behavior that appeared in Carmichael's *Manual of Child Psychology*. As befits such a function, McGraw's personal views occupy a less prominent position in this piece; rather, this is a scholarly review of research on behavioral ontogeny. Nevertheless, the piece is placed within a characteristic McGraw context: a rejection of dichotomies, such as learned versus innate and learning versus maturation, and the difficulties of defining concepts. In reviewing efforts at studying maturation, McGraw draws on research on both humans and non-human animals to consider studies in which animals are either reared in restricted environments or provided environmental enrichment of one sort or other.

Characteristically, she then turns to the issue of neural development and the correlation with behavioral development, relying heavily on the work of G.E. Coghill. McGraw was deeply concerned with problems of maturation and the structure-function relationships. She viewed the relationship between structure and function as contingent and dynamic:

> It seems fairly evident that certain structural changes take place prior to the onset of overt function; it seems equally evident that cessation of neurostructural development does not coincide with the onset of function. There is every reason to believe that when conditions are favorable function makes some contribution to further advancement in structural development of the nervous system (McGraw 1946:363).

McGraw (1946:364) concludes with a statement that will surprise none who have followed her work, noting that a "profitable approach lies in the systematic determination of the changing interrelationships between the various aspects of a growing phenomenon."

The third essay, written by McGraw in 1969, is an unpublished "Open Letter" to parents of young infants. Developmental problems are viewed in a practical context: the problem of getting women to allow their intuition to guide their child-care decisions. "Intuition," however, is not mystical, but rather is an "emulsion of sensitivity, knowledge, and feeling." This develops naturally—when women are reared in extended families. McGraw sees the mother-young interaction in dynamic systems terms. Both parent and child send and receive interacting signals. Consistency in dealing with the child does not involve adherence to a rigid set of rules, but rather depends on context—a sensitivity to the changes that occur, so that the mother's behavior will be consistent with the baby's needs at the time.

The final piece is a foreword written by McGraw in 1981 for a new series, *Advances in Infant Research*, dedicated to her. Here she touches on a variety of issues of long-standing concern to her. She opposes dichotomous concepts, sees limits to the value of data from nonhuman animals, emphasizes the implications of research for practical child rearing, and deals with issues of method. McGraw notes that growth scientists must work in an interdisciplinary context. The latter is a theme that runs throughout these pieces and an orientation that has been responsible for much of the progress in growth science since McGraw's pioneering efforts. The strongest section deals with the problem of the scientist seeking broad cultural acceptance of her research. The growth scientist must be a responsible citizen, considering the manner in which her data will be adopted and used within the culture.

THEMES AND CONTEXT

Practical Focus

McGraw's essays provide an excellent example of the melding of basic and applied research. McGraw is concerned with good, basic science. She believes strongly in the power of abstract symbols. Yet, the practical implications for the mother raising her child are never far beneath the surface. McGraw was the product of the very American functionalist-dynamic psychological tradition that was centered at the University of Chicago and Columbia University. Science was to be used for practical purposes and to be advanced through that utilitarian function. The influence of John Dewey and others at the Teachers College of Columbia University is strong. Somewhat paradoxically, however, child care has become over-intellectualized; mothers need to trust their intuition. Science can provide a guide to enable mothers to trust in themselves and reduce that anxiety over child-care decisions to which McGraw repeatedly refers.

Dichotomies

McGraw opposed the dichotomies of nature and nurture and learning and maturation, as noted in other sections of this book. She believed that genetic and environmental influences could not be neatly separated. More importantly, her quest was for an understanding of process and a nominalistic approach was of little value in that endeavor. Her views stated, for example, in her essay on maturation of behavior (see Chapter 14) are very close to what would later be termed "epigenetic" approaches.

The epigenetic approach is often presented as a solution to the nature-nurture problem, which is then labelled a "pseudoproblem." In fact, however, behavioral patterns do differ, with some suggestive of a history of natural selection and others lacking such a history. Interestingly, then, and in contrast to many later writers, McGraw retained concepts of "phylogenetic" and "ontogenetic" activities (e.g., McGraw 1946:345), which appear in various guises in her writings. Thus, "phylogenetic activities," such as swimming, appear to be under subcortical control (McGraw 1946:360). She treats this distinction in relation to the importance of the behavior for functioning as a normal human being. McGraw here is grappling with the problem that would later be treated as the ontogenetic versus phylogenetic senses of the word "innate" (e.g., Lehrman 1970). Clearly, it is the former use to which McGraw objects.

All behavioral patterns have complex ontogenies that are worthy of study and that cannot be simply dichotomized. However, some patterns appear to be innate in the phylogenetic sense that they evidence a history of selection and adaptation to the environment. Many recent authors fail to see, as did McGraw, that an epigenetic approach need not be incompatible with a distinction between ontogenetic and phylogenetic behavioral patterns. Many seem to believe that because genes and environment interact in the development of all behavior, there are no valid grounds for making distinctions among them. The problem is simplified when one recognizes that even "phylogenetic" patterns can have complex ontogenies. If one asserts that a behavioral pattern may be the product of natural selection this does not mean that it lacks a complex ontogeny and need not imply a preformationistic approach. The balance among these concepts and issues is very delicate even today but was even more so in the era in which McGraw wrote, as a strong environmentalistic emphasis was prevalent.

The Developmental Psychobiological Tradition

McGraw's work falls squarely in a long tradition of developmental psychobiology. Through much of the century, the term "psychobiology" has connoted a centrist, interdisciplinary approach in which both reductionism, on the one hand, and vitalism, on the other, are eschewed. Psychobiology has been characterized by an emphasis on dynamic processes in adapted, functioning organisms.

Perhaps the strangest lacuna in McGraw's placement in this context is her apparent unfamiliarity with developmental psychobiology as it existed near the turn of the 20th century. She viewed the Darwinian influence in psychology as leading away from an interest in developmental issues (e.g., McGraw 1981:xv). In this she erred (see Dewsbury 1984). Researchers such as Mills (1898) and Small (1899) published early studies of the development of behavior in rats, squirrels, dogs, cats, rabbits, guinea pigs, pigeons, and fowl. Leading psychologists of the time, such as G. Stanley Hall and James Mark Baldwin, had one foot planted in an evolutionary approach and the other in developmental work. Sometimes the two were not clearly distinguished (for an excellent treatment of the intertwined history of these concepts see Richards 1992). There were studies of the development of bird song and early learning. At that time "genetic psychology" implied "questions of mental evolution, development, and growth" (Baldwin 1901:410).

McGraw was correct that much of Watsonian behaviorism, particularly as it developed after World War I, was antithetical to her dynamic approach. However, the younger John B. Watson worked very much in the same tradition as McGraw. In his doctoral dissertation, Watson (1903) correlated developmental changes in the ability of rats to learn mazes with the development of their nervous systems. It is hard to understand why she did not cite this as an early example of developmental neurobiology. McGraw was correct in that there was a period after World War I when this tradition was less apparent than it had been previously. This was the period in which she received her education in psychology.

McGraw thought in terms of systems; it is no accident that she cited Bertalanffy in the 1940 essay. The emergent laws of biological systems fascinated her. The focus could be on the system at any level (McGraw 1940:82). The mother-young dyad, for example, was an integrated system based on effective reciprocal communication. Again, this is consistent with recent systems emphases (see Alberts and Gubernick 1990). It is also a reflection of her intellectual heritage. It was John Dewey (1896), in his classic critique of the reflex arc concept in psychology, who moved the field toward the study of complete functional systems and away from an emphasis on discrete stimuli and responses defined independently of the system in which they functioned.

A neurobiological approach is one important aspect of contemporary psychobiology. McGraw had a continuing interest in neural correlates of behavior that surely was a reflection of the strong influence of Frederick Tilney upon her work. Tilney developed a model in which he elaborated various stages in the phylogenesis and ontogenesis of neural structures and their relation to overt behavior (see E. Dewey 1935:16–25). Much of this work was done at a time when physiological approaches were out of favor in much of mainstream psychology; McGraw was thus swimming against some strong currents in doing in her day what seems obvious in ours.

In addition, McGraw's work is directed primarily at motor development. With the current emphasis on cognitive development in much of developmental psychology, psychobiological approaches in general, and McGraw's in particular, seem underappreciated in much of contemporary developmental psychology.

The notion of critical periods was an important part of McGraw's developmental psychobiology. She later recognized, however, that selection of that term implied a greater rigidity in development than was suggested by the data (McGraw 1985). McGraw's realization paralleled that of comparative psychologists in later decades, as emphasis shifted away from "critical" periods, as in the work of Konrad Lorenz (1973) and J.P. Scott (1962) toward one of "sensitive" periods (e.g., Bateson 1971). There are periods of maximal sensitivity to environmental input but they are neither inflexible in timing nor all-or-none in nature.

The dominant approach in recent developmental psychobiology has been toward a dynamic, epigenetic perspective (e.g., Hall and Oppenheim 1987). The ties between McGraw's approach and this recent one are obvious. A developmental psychobiological tradition was present early in the century and persists into the present. We can thus place McGraw's work within a long tradition in developmental psychobiology that stretches from workers such as Mills, Watson, and Small at the turn of the century to Oppenheim, Hall, and Alberts in recent years. The focus on dynamic processes in development has become even stronger with time.

Animal Research

McGraw opposed the direct application of principles from the study of animal learning to human infants. She repeatedly stated the theme of the inapplicability of animal studies to the study of the infant. Thus, "These learning principles were determined largely by animal studies in laboratory situations and studies of children in the classroom. There is every reason to suspect that they are not applicable to the process of growth taking place during infancy and the early years" (McGraw 1981:xxi). On the other hand, she relied on data from studies of animal development. Much of the 1946 review is concerned with studies of nonhuman animals and the animal research of G.E. Coghill was of great importance to McGraw's approach. This balance is less surprising than it might at first appear. McGraw worked during the height of the behaviorist and neo-behaviorist development of learning theories. She obviously believed that results from such studies were of little value for her quest. Characteristically, she swam against a strong current of her time. However, this should not be taken as an indictment by McGraw of the relevance of all animal research. Studies in the dynamic developmental tradition of Spalding, Herrick, Carmichael, and others revealed much about development. In sum, her criticism of behavioristic animal research should not be taken as an indictment of all animal research as applied to humans, but only of the kind of behavioristic studies that appeared dominant at the time.

McGraw (1981) attributed responsibility to Darwinian theory for two assumptions that she believed then prevalent: (1) that principles learned with animals can be simplistically applied to humans, and (2) that traits shown by the newborn are attributable to heredity. It is difficult to understand why she did not balance this indictment with the realization that the functionalist-adaptive tradition within which she worked also evolved from evolutionary theory.

Contextualization

The theme of contextualization, at various levels, may be the most significant one flowing through these essays. In the first piece, McGraw (1940) presents a very contemporary view of the socially constructed nature of concepts. A meter works as a standard of measurement only because of a series of operations, the nature of which are defined by international agreement. Concepts must be fluid, changing meaning and being refined and improved, just as one would improve a piece of apparatus. Words do not have fixed relationships to external referents, but are used in functional contexts. Even the organism can be defined in different ways for different purposes. The longitudinal method implies repeated measurements of the same observation on the same individual, but the frequency and interval depend on context. This dynamic approach to methodology is remarkable for the period.

In the second piece, the contextualization emphasized by McGraw (1946) is that inherent in development. Traits are determined by many genes and genes affect many traits. "Whatever the gene produces is contingent upon the gene combinations and the conditions under which the organism develops" (McGraw 1946:352). Thus, it is the expression of genes that is contextualized.

For McGraw (1969), in the third piece, contextualization is to be found in the dynamics of the mother-young communicative interaction. Consistency is critical. McGraw's notion of consistency however, is not constancy, but sensitivity to change. Consistency "means altering one's methods so that they are consistent within the developmental changes which have taken place in the child." There is no simple set of dicta for appropriate parental behavior. All depends on context. Mothers need to develop a sensitivity and intuition to sense these changing dynamics.

In the fourth piece, that of McGraw (1981), the most critical contextualization is that of the scientist herself. The scientist is a citizen living in a society who has the responsibility to be concerned with the uses to which her science is to be put.

McGraw must not have had a static cell in her brain. At all of these levels we see the embeddedness of her approach and a refusal to consider major questions outside of the situations in which they arose. It is this theme more than any other that is a remarkable aspect of McGraw's thinking.

References

Alberts, J.R., and D.J. Gubernick. 1990. "Functional Organization of Dyadic and Triadic Parent-Offspring Systems." In *Mammalian Parenting: Biochemical, Neurobiological, and Behavioral Determinants,* eds. N A. Krasnegor and R.S. Bridges. New York: Oxford.

Baldwin, J.M. 1901. *Dictionary of Philosophy and Psychology.* New York: Macmillan.

Bateson, P.P.G. 1971. "Imprinting." In *The Ontogeny of Vertebrate Behavior,* ed. H. Moltz. New York: Academic Press.

Dewey, E. 1935. *Behavior Development in Infants: A Survey of the Literature on Prenatal and Postnatal Activity 1920-1934.* New York: Columbia University Press. Reprint edition New York: Arno Press, 1972.

Dewey, J. 1896. "The Reflex Arc Concept in Psychology." *Psychological Review* 33:357–370.

Dewsbury, D A. 1984. *Comparative Psychology in the Twentieth Century.* Stroudsburg, PA: Van Nostrand Reinhold.

_____. 1991. "Psychobiology." *American Psychologist* 46:198–205.

Hall, W.G., and R.W. Oppenheim. 1987. "Developmental Psychobiology: Prenatal, Perinatal, and Early Postnatal Aspects of Behavioral Development." *Annual Review of Psychology* 38:91–128.

Lehrman, D.S. 1970. "Semantic and Conceptual Issues in the Nature-Nurture Problem." In *Development and Evolution of Behavior,* eds. L.R. Aronson, E. Tobach, D.S. Lehrman, and J.S. Rosenblatt. San Francisco: Freeman.

Lorenz, K. 1973. *Motivation of Human and Animal Behavior: An Ethological View.* New York: Von Nostrand Reinhold.

McGraw, M.B. 1940. "Basic Concepts and Procedures in a Study of Behavior Development." *Psychological Review* 47:79–89.

_____. 1946. "Maturation of Behavior." In *Manual of Child Psychology,* ed. L.Carmichael. New York: Wiley.

_____. 1969. "An Open Letter to Parents of Young Infants." Unpublished manuscript (August 2). *Myrtle McGraw Papers,* Special Collections, Millbank Memorial Library, Teachers College, Columbia University, New York.

_____. 1981. "Challenges for Students of Infancy." In *Advances in Infancy Research Vol. 1.* L.P. Lipsitt and C. Rovee-Collier, eds. Norwood, NJ: Ablex.

_____. 1985. "Professional and Personal Blunders in Child Development Research." *Psychological Record* 35:165–170.

Mills, W. 1898. *The Nature and Development of Animal Intelligence.* New York: Macmillan.

Richards, R.J. 1992. *The Meaning of Evolution.* Chicago: University of Chicago Press.

Scott, J.P. 1962. "Critical Periods in Behavioral Development." *Science* 138:949–958.

Small, W.S. 1899. "Notes on the Psychic Development of the Young White Rat." *American Journal of Psychology* 11:80–100.

Watson, J.B. 1903. *Animal Education.* Chicago: University of Chicago Press.

❧ 13 ❧

Basic Concepts and Procedures in a Study of Behavior Development

THE ROLE OF CONCEPTS
IN SCIENTIFIC PROCEDURE

The time can still be recalled when students preparing for experimental work, especially in the biological sciences, were exhorted to "stick to facts," or "let the facts speak for themselves." Young investigators were made to feel that the business of science was much like that of exploration—the world of the unknown was there and it was almost the sole duty of the scientist to unearth these hidden facts. Once brought to light such facts would stand boldly evident as part of our body of scientific knowledge.

But those times have changed. Certainly the advancing guard of investigators today realize that concepts pertaining to the subject-matter and to the methods employed in manipulating or intellectualizing data are tools just as significant to experimentation and research as are the instruments of measurement or the actual facts obtained. Concepts of the investigator and those which have accumulated about the subject-matter are salient factors in the process of investigation, whether they are recognized as such or not, and they strongly influence the manner in which facts are obtained, the interpretation of the data, and the organization of knowledge about the data. It would therefore seem advisable to admit the role of personal and social concepts in scientific undertakings and to work toward their improvement and refinement as deliberately as one would set about improving a piece of apparatus. It should also be recognized that concepts and methods as tools of inquiry alter with time. Often these conceptual alterations are quite haphazard and uncultivated; sometimes the usefulness of a particular concept is thereby utterly destroyed. A concept which was quite definite and precise in its origin may in the course of time acquire such a cargo of meanings that its original

Reprinted from *Psychological Review*, 1940, 47(1):79–89, by permission of Mitzi Wertheim.

identity is lost. Others are vague and loose in their inception but attain specificity and definition as time goes on.

It is sometimes possible by arbitrary and deliberate demarcations to save a useful concept the unhappy lot of becoming a symbolic football in the game of scientific polemics. The concept of meter, for example, is one which has attained precision by virtue of national and international agreement. There is no specific entity called meter. The concept of "meter" is often considered as a name for something concrete in existence, because it has existed for so long in our scientific thoughtways that its origin and actual import are frequently overlooked. Those investigators who are dealing with phenomena and relations for which no useful units of measurement, such as meter, are available would do well to pause and consider the manner in which such concepts of measurement in the physical sciences gained their present degree of specificity.

> The standard meter is a bar of platinum kept under as constant conditions of temperature and pressure as possible in the city of Paris. But if that were the whole story, the word *meter* would not have the connection with measuring it actually has. By itself the bar is just a particular bar and nothing else; it is neither a standard of measurement nor is itself measured. It is a measure of length because (1) all other rods of a meter's length in use anywhere in the world may be checked by being matched against it, and (2) because, and *only* because, these other rods are themselves used in matching still *other things*. It is just as true that the length of the bar of platinum (or any other measuring rod) is determined by its application in measuring cloth, walls, sides of fields, etc., as that the length of the latter is determined by comparison with it (Dewey 1938:214).

It should be borne in mind that the now specific concept "meter" developed out of vague descriptive analysis of the relations between things in space and that it arrived at and retains its precise meaning by virtue of international agreement. Furthermore, in its present use it represents not single units but a series of operations which form the basis for comparing relations between things in existence. This illustration is offered because a little retrospection on the origin and development of some of these precise units of measurement in the physical sciences would no doubt be of aid in bringing about greater precision in concepts of biological phenomena, especially in studies of behavior development where one is often dealing with a constantly changing phenomenon which does not lend itself readily to quantitative determinations.

CONCEPT OF "ORGANISM" AS REPRESENTING A FIELD OF INQUIRY

But before becoming too concerned over units of measurement or quantitative determinations, investigators in the field of behavior development should first of all consider the nature or quality of the phenomenon being investigated. For the moment may we say that any study of behavior development embraces an analysis of changes taking place within a living organism. At the mention of the word "organism"—to say nothing of "living"—we are beset with difficulties, and are immediately confronted with a concept in dire need of refinement. Recent literature is replete with discussion of the "organism as a whole" and it is again and again emphasized that the functioning of the total organism is something quite different from the assembling of parts or elements of which it is composed. But it can safely be said that the term "organism" has not been subjected to a rigorous definition. Certainly the term "organism" implies organization and organization is the essence of living matter. Bertalanffy emphasizes this point when he says:

> Since the fundamental character of the living thing is its organization, the customary investigation of the single parts and processes, even the most thorough physio-chemical analysis, cannot provide a complete explanation of the vital phenomenon. This (type of) investigation gives us no information about the coordination of the parts and processes in the complicated system of the living whole which constitutes the essential " nature " of the organism and by which the reactions of the organism are distinguished from those of the test tube. But no reason has been brought forward for supposing that the organization of the parts and the mutual adjustments of the vital processes cannot be treated as scientific problems. Thus, the chief task of biology must be to discover the laws of biological systems to which the ingredient parts and processes are subordinate. *We regard this as a fundamental problem of modern biology* (Bertalanffy 1933:64–65).

Yet granting that an organism represents the organization of living systems, we still do not know, for experimental purposes, just what constitutes the boundaries or the scope of any particular organism. As the term is used, it seems to refer in most cases to the scope or field of inquiry rather than to any peculiarly identifying characteristic. It may be easy enough to identify a single cell floating around in plasma as an individual organism and one may undertake a study of the organization of nucleus and cytoplasm therein contained, but when one begins a study of the behavior of mammals, the business of identifying tangible boundaries of an individual organism is infinitely more complex.

Would it not be more practical, therefore, to accept an operational definition of the organism? That is to say, the phenomenon to be observed is "organization," and the limits or the field of observation, are prescribed in accordance with each investigation. For some purposes this field, or scope of observation, may be a single cell, for others an individual as bounded by epidermis, or it may be a functional system as it operates within the epidermis, and for still other purposes it may be the universe. Actually there are no sharp lines which clearly demarcate an individual, or an organism, except as they are experimentally or operationally determined. Socially speaking, an individual represents merely nodal points of activity within a society and for some purposes the whole society may constitute the experimental organism. In much the same way that arbitrary limits can be prescribed around nodal points within a society to signify a particular individual, so may arbitrary limits, for experimental and observational reasons, be set around nodal points of a psycho-motor activity. These nodal points of a psycho-motor expression may be called a behavior pattern, or an experimental organism, so long as it is the organization of neuromuscular movements which concerns the investigator. Once the limits of observation have been deliberately established the changes in the organization of the activity over a period of time would comprise a study of behavior development.

Organization, then, is a quality varying in amount and common to all matter and energies. It may strain the imagination but certainly it is not altogether inconceivable that the concept of organization could be so defined that a unit of measurement could be applied which would indicate degrees of organization without regard to the particulars of the organ-ism, similar to the way in which a mercury thermometer may be used to indicate changes in the temperature of things which are qualitatively quite different. It was not until temperature was conceived as a quality permeating substances which were qualitatively unlike, that a device for obtaining symbolic measures of temperature in varying substances was feasible.

Degrees, as indicated on a mercury thermometer, are not measures of temperature directly, but are merely symbols representing a quality which has been intellectually abstracted from the substance in which it exists (Dewey 1938:214). It is not necessary to extract the temperature of a particular substance in order to match it against the temperature of another. The parallel is obvious; it would not be necessary to dissect "organization" from the organism in order to study its process of change, provided some proper symbolic indicator could be devised and applied as a measure of organization. Construction of such a symbolic system must await refinement in the concept of organization and its meaning with respect to living organisms.

In any study of behavior development a necessary procedure at the outset is the determination of the scope of inquiry. This field of inquiry is arbitrarily determined by the investigator and the nature of the problem. In studying the development of reaching-prehensile behavior of the infant, for example, the object in the field of vision is just as much an integral part in the organization of the behavior as are the arms, fingers and eyes of the baby. And if one sets out to study the organization of this activity, the bottle for which the baby extends his arms and fingers is a part of the behavior function as well as the neuromuscular movements which the baby makes. One manipulates arms and fingers quite differently when picking up a bowl of water from the way one does when trying to catch a fly. In that sense the object determines the configuration of neuromuscular movements, and as such it might be considered "an organizer" of behavior. But the manner in which a baby would pick up a bowl or catch a fly is still different from the way in which the adult would do it. It is these changes in manner of performance which reflect different degrees of organization in a given behavior activity. The boundaries of inquiry, that is to say, the experimental organism, having been deliberately prescribed, the next procedure involves an analysis of the changing degrees of organization within these boundaries.

THE "LONGITUDINAL METHOD" OF COLLECTING DEVELOPMENTAL DATA

Admission of the word "development," or changes in behavior, imposes upon the investigator the necessity of observing the same or comparable observational fields from time to time. It is the realization of this necessity which has evoked the current emphasis upon "longitudinal" method in the biological sciences, especially in the field of child development. Along with this emphasis considerable confusion has accrued as to what is actually involved in a longitudinal method of investigation.

Dr. Palmer (1937) in discussing this state of confusion with regard to the longitudinal method said:

> In a general way most workers in the field of growth and development would agree that it is possible to classify data dichotomously, one type being known as longitudinal and another type being known as cross-sectional. Agreement may also be expected with respect to the principal characteristic which differentiates these two classes of data. Thus it is generally recognized that cross-sectional observations are those made on different individuals while longitudinal observations are those made on the same individuals. All of this, however, scarcely furnishes even working definitions and there seems to be considerable difference of opinion regarding more explicit definitions.
>
> According to one concept which more or less expresses the views of some workers, the matter is handled by defining longitudinal data as follows: Longitudinal data consist of a long series of repeated measurements of the same characteristic observed on the same individuals. The broader aspects of this definition appear to be reasonably clear. For example, it is generally agreed that records of the weights of a group of children, each weighed on their birthdays for a period of years, constitute longitudinal growth data. It is clear in this case that the same characteristic (weight) must be measured several times. It is equally clear that successive weighings must be made of the same children.
>
> The part of the above definition which causes difficulties concerns the idea that 'a long series of repeated observations' are required if data are to be called longitudinal. This idea leads, I believe, to difficulties as regards a clear concept of what longitudinal data are, as regards the usefulness of longitudinal data and also as regards methods of analysis of such data.

What is a 'long series' and what are 'repeated observations?' The implication is clear that some interval of time should or must elapse between repeated observations. It is implied also that many observations must be made. The question of how long the interval of time must or should be, or of how many repeated observations must or should be taken, is not explicitly stated so far as I know.

In an attempt to clarify these points, I should like to give and defend another definition of longitudinal data. This definition is as follows: Any datum consisting of two or more observations of the same characteristic observed on the same individual shall be considered longitudinal.

In amplifying this statement, it may be emphasized that the interval between successive measurements may be any time whatsoever. In actual practice, the only limiting factor which is placed on the interval of time between measurements is that the interval have meaning in the particular problem under investigation.

Palmer's definition of longitudinal data is, in a sense, a refinement of our loose concept, but in another sense it is too broad to illuminate or spot the essence of a longitudinal study. According to his definition weight measured at birth and again when the child is ten years old would yield longitudinal data. Such data would not, however, reveal the way in which the weight increment occurred. Of course it is the manner in which data are handled and interpreted as well as the frequency with which they are collected that indicates their longitudinal merits. But it would be necessary to have at least three observations on any phenomenon in order to show its course of growth, unless we are going to admit that all development is in the order of a straight line.

It is in studies of growth and development that the longitudinal method gained eminence. Growth and development are concerned with change—change in form or configuration, or change in magnitude. Any truly longitudinal study must be conducted in such a way as to indicate the nature of the changes taking place in the observed phenomenon. It would seem, therefore, that a longitudinal study would be comprised of any series of observations on a changing phenomenon taken successively from the moment of inception until the changing characteristic attains stability or decline. The moment of inception and the state of stability or decline should be arbitrarily or operationally defined in terms applicable to the experimental procedures. It must, furthermore, be emphasized that the data should be taken in such a way that they can be handled intellectually and mathematically so as to take advantage of their longitudinal qualities, that is to say, handled in a way to show not only the fact of change but also the trend, velocity, and nature of such changes.

The crux of the difficulty in collecting longitudinal data seems to hinge primarily upon the interval between observations. Palmer has suggested that the interval between measurements may be "any time whatsoever" so long as the interval has meaning with respect to the problem being investigated. In most of the longitudinal studies published in the field of child development, the general assumption has been that the intervals between observations are adequate provided they embrace all changes of any detectable magnitude. This assumption certainly has logical meaning as it bears directly upon the problem being investigated. On the other hand, if it is fully realized that it is the way in which the data are handled which determines the longitudinal merit of a study, and not the fact that data are collected in serial order, then it will be also recognized that the interval between observations is determined by the demands imposed upon the data for adequate mathematical and intellectual analysis; if one is dealing with a characteristic for which numerical values are obtainable and if the approximate rate of change in that particular characteristic is generally known, then the interval between measurements may be calculated at frequencies just sufficient to embrace all measurable changes. For example, it would be wasteful and laborious to measure the degree of bone calcification in the wrist of the same child every day. In the first place, the degree of change from day to day is incommensurable according to our present methods, and present methods of measurement do yield definite numerical values which can be manipulated for the purpose of indicating velocities, accelerations, regressions, and trends of development in that particular characteristic. It is wrong to make daily records on a characteristic of this type only because it is wasteful. On the other hand, if one is dealing with a changing characteristic for which no numerical or distinctly quantitative values obtain, or if one is dealing with a characteristic the general trend and developmental rate of which is not known, then the data must be accumulated at great frequency, even when no significant changes are taking place. The increased frequency of observations makes it possible to reduce the data to some symbolic form so that they can be handled mathematically or intellectually in a way that would not be possible with the original facts. Certain types of data may be recorded longitudinally at sufficient intervals to include all significant changes in the developing characteristic, and yet the data will remain utterly useless for symbolic or intellectual manipulation. For example, it is possible to record on a particular baby that at the end of the first post-natal month he showed the typical Moro reflex (McGraw 1937) in response to a slap on the bed, at the end of the second month the reflex was still present though there was less bowing of the arms over the chest, and that at the end of the third month the gross reflex had diminished and the child gave only a general body jerk to this type of stimulus. In this case the data are longitudinal, recorded serially upon the same child, and the records embrace all the significant changes which take place in that particular characteristic. But intellectually there is very little one can do with such data other than repeat the descriptive statements of which the data are composed.

It has been repeatedly emphasized in this discussion that it is not so much the growing phenomenon per se but rather the mathematical demands upon the data which determine the interval and limits of observations. At the same time it should be recognized that the acceptance of mathematical demands instead of the status of existing phenomenon as the criterion of observational procedures is entirely justified because symbolic data, once they are organized into a meaningful system, can be manipulated on their own account so that relations and meanings can be educed from these manipulations which perhaps could not have been detected in the direct data. It is indeed, in this sense that mathematics becomes the true handmaiden of scientific inquiry, not merely an elegant gown in which an investigation is dressed for public appearance.

One does not need to look to science in order to appreciate the advantage of an ordered symbolic system in dealing with an actual situation. Any motorist who struggled with the old descriptive Blue Books, less than a quarter of a century ago, knows the value of modern road maps and the numbered highways. It is due not merely to the fact that highways have been expanded and improved but also to the ordering of highway information to a symbolic system that present-day motoring can be carried on with facility. Again it should be emphasized that the reason for ordering observational or direct data in a symbolic system is because the symbols can be manipulated within their own system, independent of the actual phenomenon.

In emphasizing the influence of a symbolic order for determining the manner and frequency at which observations or measurements of a growing phenomenon should be made, the point should also be stressed that the original data may and often should lead to the formation of a new symbolic order. Mathematical systems are not finished to the extent that additional symbolic orders are precluded. Woodger (1937) and Lewin (1936) are both working on new symbolic systems for the ordering of biological data, and while the writer is not qualified to evaluate these methods, the fact that work is being done in that direction is reason for encouragement.

❧ 14 ❧

Maturation of Behavior

INTRODUCTION

Early interest in structure and function was concerned not directly with behavior but with specific organs and their use. It harks back to the days when
Lamarck (1773), emphasizing the development of soma through exercise,
formulated a theory which has come to be known as the inheritance of acquired traits. Lamarck not only contended that individuals improved in
function through exercise of somatic muscles but that it was the protracted
use of muscle groups which effected the creation of new organs and new species. Amid these propositions are found quaint ideas to the effect that the gi-
raffe gained his long neck by reaching into the trees for food; that the snake
lost his legs by creeping through narrow crevices. When one considers the
abundance of evidence in everyday life showing the increase in muscle
strength through exercise it is no wonder that the Lamarckian theory gained
credence. As an explanation for the mechanism of heredity, however, the
theory was dealt a severe blow when Weismann (1889) pointed out the distinction between *germ plasm* and *somatoplasm,* the continuity of germ
plasm from generation to generation, and its comparative insulation from
other tissues of the body. Weismann's contentions also served to crystallize
the dichotomy of heredity and environment.

Following the work of Weismann and the resurrection of Mendel's (see
Castle 1921) papers in 1900, investigators in the field of behavior as well as
morphology were laboring under the conviction that the nucleus of the germ
plasm was the sole agent of hereditary factors, and that the germ was impervious to modification by environmental stimuli. As a result of this dichotomy the bulk of psychological literature during the early part of the twentieth century was concerned with the determination of those aspects of human
behavior which are hereditary in origin as contrasted with other types of be-

havior which are acquired individually. Textbooks (Angell 1908; McDougall 1914; Thorndike 1919; Perrin and Klein 1926) of the time customarily classified behavior into (1) reflexes, (2) instincts, and (3) acquired traits. Such classifications were based upon a consideration of the complexity, variability, origin, and modifiability of the activity concerned. The labeling of human behaviors in categories of this order was for the sole purpose of facilitating ratiocination about the nature of man. The vast amount of literature accumulated within a decade on the subject of instincts and acquired characteristics bears testimony to the assertion that the system defeated its own end. Soon the various compilations of instincts became so diverse that they were intellectually unwieldy.

The influence of these controversies, however, was at least twofold. On the one hand they served to emphasize the distinction between innate and acquired characteristics, and on the other hand they stimulated extensive experimentation for the purpose of determining those traits which are predominantly hereditary in origin. The complexities which arose from such categories made it evident, however, that the conceptual framework in which the workers of the time were operating was in need of revision.

The revolution came with the birth of *behaviorism,* when instincts and consciousness were swept aside. Actually, the foundation of the heredity-environment dichotomy was left standing and a new structure was built on the old foundation. The essential difference between the new and the old structure was a matter of degree. The behaviorists (Watson 1919; Weiss 1929) readily admitted that there was a distinct difference in the development and function of traits transmitted through the germ plasm and those individually acquired, but they greatly restricted the number of characteristics which could be so transmitted. As a result of this shift in emphasis they have gone down in history as ardent environmentalists. In 1914 Watson, the chief protagonist of behaviorism, asserted:

> The student of behavior has come to look upon instinct as a combination of congenital responses unfolding serially under appropriate stimulation; the series as a whole may be 'adaptive' in character (always adaptive from the Darwinian standpoint) or it may be wholly lacking in adaptiveness. Each element in the combination may be looked upon as a reflex. An instinct is thus a series of concatenated reflexes. The order of the unfolding of the separate elements is a strictly heritable character. Instincts are thus rightly said to be phylogenetic modes of response (as contrasted with habit, which is acquired during the lifetime of the individual).

In making this statement, Watson not only reduced instincts to a series of reflexes, but also suggested a new theory of behavior development. The theory might be reduced to simple terms in the statement that the hereditary endowment of the individual is limited and the hereditary units of behavior can be identified by structural or physiological correlates. These innate units of behavior are conveniently termed "reflexes" and by a system of concatenation, association, conditioning, etc., more complex activities are constructed from these simple behavior units. There were several existing conditions which helped the theory to flame. In the first place the popular scientific mood at the time was to reduce each phenomenon to its smallest unit; the simple reflex was the psychologist's answer to this urge. Furthermore, the work of Pavlov (1927) had suggested the means whereby the simple reflex units of behavior might become complex constructs; so the business of the psychologist was to determine how, by a process of "conditioning," and "reconditioning," extraordinarily complex behavior mechanisms could be effected. Genetics (Lock 1916; Conklin 1916; Castle 1921; Morgan 1932) at the time not only burdened the ultramicroscopic gene with the hereditary load but also postulated that within the germ plasm there were definite determinants for every heritable character. This theory has come to be known as the inheritance of unit characters. Naturally if the ultramicroscopic gene was to carry the load for every heritable trait, the limitation of heritable characters to a minimum seemed more reasonable. Furthermore, the theory of the environmentalists gained credence because of its simplicity, and it achieved great popularity because of its hopeful outlook.

On the other hand, theories of behaviorists stimulated other lines of approach in the search for a better understanding of the complexities of behavior development. Their emphasis upon physiological correlates of behavior certainly enhanced, though it would not be claimed to have activated, the more recent investigations concerning the correlation of neurostructural development and behavior. Indeed, so extensive has the work been along these lines that the phrase "structure and function" has practically lost its original connotation of organ and use and has come to refer essentially to the relationship between cytological structures of the nervous system and behavior. Moreover, the behaviorist group, and particularly Watson, can justly lay claim to some credit for the current vogue of genetic psychology, especially child psychology, even though the early infant studies by Watson (1919) were primarily to determine the repertoire of innate endowment. But as soon as attention shifted from inventories of behavior traits to developmental changes in activities it became evident that the mechanistic theory, ascribing complex behavior to a concatenation of simple reflex units, was an inadequate explanation of the intricate process involved in development. Hence another term and another dichotomy have gained prominence in more recent literature. The term is *maturation,* and the dichotomy is *maturation and learning.*

DEFINITIONS OF MATURATION

Opposing the rigidity of the environmentalists' viewpoint, Gesell (1929) has been particularly active in popularizing the term maturation through the literature on child psychology. In its original application as a scientific term the word was used by geneticists and embryologists to denote that period of development prior to fertilization which converts an immature germ cell into a mature one. The significant change that takes place during this phase is the reduction of chromosomes, so that the mature germ cells (ovum or sperm) have only half as many chromosomes as the immature germ cell or other body cells. (See Waddington 1936:31; Jennings 1935:46). Clearly the original meaning of the term maturation has specific reference to the organization of chromosomes in the germ cells, although it has acquired a more general application in biological as well as psychological literature.

In his observation of infant behavior development, Gesell (1925) had repeatedly observed phenomena which could not be adequately explained in terms of the conditioning theories of the behaviorists in vogue at the time. Gesell (1933) felt that such theories

> do not give due recognition to the inner checks which set metes and bounds to the area of conditioning. ... Growth is a process so intricate and so sensitive that there must be powerful stabilizing factors, intrinsic rather than extrinsic, which preserve the balance of the total pattern and the direction of the growth trend. Maturation is, in a sense, a name for this regulatory mechanism.

It is not possible to identify Gesell's use of the term with any definite physiological process, but he justifies his broader application of the term, since, as he (1933) says:

> It must be remembered that the genes do not find lodgment in the reproductive cells only but in all the somatic cells. Into each cell of each tissue and of every bodily organ go appropriate subdivisions of both paternal and maternal genes. These ancestral genes are found in every neuron. They produce, as well, the hormones which secondarily regulate development at all ages, prenatal and postnatal. It is these genes which are the focal sources of directive and constructive energy. They interact with the cytoplasm, which is always influenced by intracellular and extra-cellular environment; but the primary physiological factor in this interaction traces to the gene.

It is evident that Gesell does not use the term *maturation* to indicate any particular rearrangement or reduction process of the chromosomes within the somatic cells as the individual grows. These quotations in isolation sound as if he considered the genes sitting, like censors, with in the organism telling environment when to stop. Actually, such an interpretation would be most unfair. Gesell was seeking an expression for those phenomena which develop in an orderly fashion without direct influence of known external stimuli. The older term *instinct,* having fallen into scientific disgrace, was a feeble instrument with which to combat the tenets of the radical environmentalists. In a real sense, however, the sponsors of the maturation theory have taken up the battlefront abandoned by the contenders for the instinct theory and are reasserting the role of biological inheritance in the development of behavior performances. In that sense they are confronted with the same problems, and the criteria and methods employed in experimentation are not unlike those used in the study of instinct.

In the first place, they are concerned with the appearance of particular abilities without the benefit of practice; they are concerned with the sudden appearance of new behavior items;. with the consistency of behavior patterns of different subjects of the same species with an orderly sequence in the manifestation of different patterns; and, finally, with the gradual saltatory course of growth. The facts that most babies are able to reach for an object within the visual field before they can maintain the sitting position, that they sit up before they stand, and creep before they walk are presented as evidence that genetic constituency controls the order of development. Shirley (1931b; 1933a; 1933b; 1933c) particularly has emphasized the consistency in a sequential order of development and the sudden emergence of new behavior items as criteria of maturational development. Some of the points at issue in discussions of ontogenetic development are reminiscent of similar issues raised in controversies of the evolutionary process. For example, the question as to whether development is gradual or saltatory (Shirley 1931c) harks back to old disputes over continuous and discontinuous variation.

Other writers apply the term more loosely as referring to any phenomenon in the process of completion without reference to physiological or genetic correlates.

Maturation may be defined precisely as the progress of an immature organism toward a mature, or terminal, state—a process produced by constant forces acting under constant conditions. The organism itself may be either individual or social, and the mature state may be either that of the organism as a whole or that of any of its parts or functions (Courtis 1935).

The term had not found its way into psychological textbooks of twenty years ago, but the complexity of the concept is reflected in the definition proposed in a recent textbook (Boring, Langfeld, and Weld 1939): "Maturation consists in structural changes which are mainly hereditary—i.e., which have their origin in the chromosomes of the fertilized ovum—which are also in part a product of the interaction of the organism with its environment."

Indeed, the concept is so difficult to manipulate that some writers have taken the position that any attempt to separate the maturational and learning processes only adds to intellectual confusion. Carmichael (1933) emphasizes that from the moment growth has begun in the fertilized ovum until senescence or death, development consists in the alteration of existing structures and functions in an organism living in a continually changing environment. That is, it is not possible save for pragmatic reasons to say at any point that growth has stopped and learning has begun, but that the environment plays a part in all 'maturation' and maturation plays a part in all learning.

Gesell, Thompson and Amatruda (1934) admit that it is not desirable to belabor the distinction, but assert that scientifically it is a desirable distinction and that it is analytically feasible. After all, the only justification for making classifications of any order is that they facilitate intellectualizing and manipulating concepts about phenomena. Marquis (1930) takes exception to what he calls the "convergence theory" represented by Carmichael and others, since "it practically precludes the possibility of experimental investigation; because it dismisses the problem of innate behavior." He further points out what he believes to be a valid distinction between a maturational process and learning:

> Both processes represent an interaction of organism and environment, but learning is distinguished from maturation by this fact: It represents a modification of the organismic pattern in response to specific stimuli present in the external environment at the time of the modification. Maturation, on the other hand, is a modification of the organismic pattern in response to stimuli present in the inter-cellular and intra-cellular environment which at the given moment are independent of external influences.

He also emphasized that intracellular and intercellular environments are hereditary components, derived from the parent in a manner comparable to the inheritance of the germ plasm.

Marquis' definition at first glance appears to denote valid and workable distinctions. On examination, however, the definition is confusing and calls for a clarification of the terms *organism* and *internal* and *external environment*. (McGraw 1940a). This confusion becomes more evident when one ponders the question: To what are these environments internal or external? Furthermore, "specific stimuli present in the external environment at the time of the modification" do not envisage those past experiences which, though ineffectual at the time, become functionally operative only after subsequent development of another order.

Later, even Carmichael (1936) took the position that *maturation* and *learning* could be given valid meaning although physiologically they were not wholly inseparable. At this time he stated:

> The development of 'our knowledge of external things,' and behavior change associated with such development, is dependent upon the fact that certain receptor mechanisms are specialized to be stimulated by one sort and only one sort of energy. ... From this point of view, the development of adaptive behavior

associated with receptor stimulation in ontogeny becomes the account of the gradual evolution and change in the organism of anatomically specialized receptor neuromotor systems of such a sort that their so-called resting activity may easily be affected only by very definite changes of a limited kind in the external world of stimuli.

He defines external environment as "the sum total of all the physico-chemical energies releasing neural activity as a result of action on *exteroceptors*" and internal environment as the "sum of energy changes acting on *interoceptors* and *proprioceptors,* even though they involve deforming pressures or chemical or electrical stimulation brought about by the body's own active movement." He suggests that the term *learning,* in contrast to maturation, "may be used to characterize the development of responses which can be demonstrated to be changed because of certain antecedent stimulus-released processes." Despite his admission that a valid distinction may be drawn between the two concepts, he still contends that such words add little to the study of behavior development and as a final proposal he suggests that: "the facts of developmental change in behavior can best be represented as empirical curves and that generalizations derived from such curves, together with the properties which they represent, would not be appreciably enriched by labelling them under a less specific nomenclature."

Digressing for the time being from the problem of genetics in the process of maturation, let us consider the practical issues involved in ontogenetic development. Specifically the issues are two, viz.: (1) What are the physiological changes in neurosomatic tissues—structural changes and changes in chemical and electrical function—which correlate with developmental changes in overt behavior? (2) In what way and to what extent does the change (increase or decrease) in practice or exercise of a function accelerate or retard its development? As a corollary of this question, one is prompted to speculate whether or not the practice affects the changes in neurosomatic structures, or whether the modification of overt behavior must await organic changes in the neuromuscular system. Evidence brought to bear on these subjects is generally in the nature of (1) investigations which have in some way attempted to alter or measure the factor of practice and to determine practice-effect by measurement of overt behavior and (2) investigations which have aimed at altering the neural organization or otherwise at determining the neural status at the onset of behavior function.

METHODS OF INVESTIGATION

Practice or Exercise the Experimental Variable

In general, those investigations which have operated within a framework of overt behavior only may, for convenience, be considered in the following categories: (1) those which base their criteria of maturation primarily upon the accuracy or precision of a given function at the time of or immediately after its emergence; (2) experiments where function has been experimentally restricted beyond the expected period of its manifestation; (3) experiments where practice or repetition of performance has been experimentally induced for the purpose of determining the effect of practice *per se* upon performance; and (4) experimental situations where the effects of both repetition and restriction of performance might be ascertained with respect to the same developing phenomenon.

Some of the early familiar studies on the subject of instincts bear repetition lest they be forgotten and their contributions in the study of maturation be lost. That the concepts of *instincts* and *maturation* are kindred in nature is evidenced by the frequent similarity in experimental methods employed and the criteria proposed as a basis for analysis. So close is their relationship that Witty and Lehman (1933) contend that the term maturation refers merely to phenomena previously known as *delayed instincts*. A familiar criterion of instincts was the sudden appearance of a particular behavior pattern and the precision of its function. These same arguments are set forth in current distinctions between maturation and learning. In 1911, Breed pointed out that the pecking of chicks is not a function which burst forth in the full bloom of perfection. Even Shirley (1931a; 1933a; 1933b; 1933c), who lays great stress upon the sudden manifestation of new motor activities in the growing infant, does not claim that they emerge at a peak of perfection. Certainly it seems that in the higher organism no adaptive behavior function is at its optimum from the moment of inception. For that reason the perfection or efficiency criterion alone seems to be an inadequate basis upon which to determine innate phenomena or the fundamentals of the maturational process.

The technique of restricting function beyond the inception period has been more fruitful. As early as 1873 Spalding had employed a method of hooding young chicks to prevent their seeing or pecking at food for several days. Spalding made many observations on the behavior of chicks and other fowls, and among some of his interesting observations he noted that young chicks who had been hooded and kept from the hen for a day or two would, when released, run immediately to her in response to her call, but if the chicken had been kept from the mother for a period of ten days it could not be induced to follow her. The moment of "ripeness" for that particular behavior had apparently been exceeded, and presumably some other method would have to be introduced in order for the chick to achieve a type of behavior which would have been simple a few days earlier. Spalding (1875) also used the restraint method in flight prevention of young swallows beyond the time when they would ordinarily begin to fly. He reports varying degrees of efficiency with which different birds made their first flight, although they all exhibited flight movements.

Recently Dennis (1941) used buzzards in a repetition of Spalding flight experiments. Dennis caged two young birds just beginning to feather and prior to the normal manifestation of flight behavior. Ten weeks later, after unrestrained immature buzzards had been soaring with adult skill, the flight behavior of the caged birds was examined. At this time the birds sometimes spread their wings as they ran and in so doing on one or two occasions raised themselves off the ground for a distance of two feet only. Even when the birds were placed on the perch they not only exhibited poor balancing ability but fell, instead of trying to fly, from the perch to the ground. Experimental observations were continued on these birds for a period of several weeks. The evidence is strong that the impaired flight ability of the birds was a result of their prolonged captivity.

Shepard and Breed (1913) fed chicks artificially while they were kept in a dark room and prevented them from pecking at food for periods of three, four, or five days. On testing, they found that it required about two days for the chicks to gain normal efficiency in pecking and swallowing. The rapid improvement which occurred during those two days they attribute to practice, and subsequent improvement to maturation.

Yerkes and Bloomfield (1910) utilized the prevention method in studying the tendency for kittens to catch mice. They observed that "suddenly" during the second month the behavior of the kittens changed so materially when the mouse was introduced into the cage as to justify the conclusion that kittens instinctively kill mice. They pointed out, somewhat incidentally, that this type of behavior was increasingly difficult to evoke as the animal grew older. This observation, along with the responses of Spalding's chickens to the mother hen, suggests that there may be a favorable maturational status for the elicitation of any particular type of behavior, but if it is not afforded external stimulation the original tendencies become less sensitive. Attention is called to these incidental observations because they were, in both instances, passed off as of small consequence, but, as a matter of fact, they may be of greater significance in elucidating the meaning of maturation than were the central data of these studies.

Kuo (1930) did a more elaborate study on the rat-killing tendencies of kittens, one aspect of which was to keep 20 kittens in isolation so that they would have no chance to observe rat-killing behavior of other cats. Of these 20, 9 did at some time during the first four months of life kill the rat, and 11 did not. Kuo offers no satisfactory explanation for the predatory behavior of the 9 kittens except to disclaim the usefulness of a term such as instinct to identify such a phenomenon.

Carmichael (1926) showed considerable ingenuity in his use of the restriction method in studying the swimming movements of the young frog and salamander. Taking embryos at an early stage of development before any bodily movements appeared, he divided them into two groups. One group he allowed to develop in ordinary tap water; the other group he placed in a solution of chloretone, of just sufficient concentration to anesthetize the animal but not concentrated enough to impair its physical growth. The larve in tap water grew more rapidly in size than did the experimental group, but there was no distinction in organ differentiation. Bodily movement in response to external stimulation was absent in the experimental group so long as they were in the chloretone solution. After the control embryos had begun to display swimming movements, the experimental ones were taken from the chloretone solution and placed in ordinary tap water. It was found that, on the average, within less than twelve minutes they began to respond to external stimulation and within thirty minutes these previously drugged embryos were engaging in swimming movement. The salamanders swam so well within that time that it was only with difficulty that they could be distinguished from the controls. Subsequently, Carmichael followed up the original study by checking the influence of the anesthesia , or the time of recovery. He re-anesthetized the animals after they had been for a period of twenty-four hours in tap water. On removing them from the chloretone solution the second time he found that the time interval between removal from the chloretone solution and the onset of swimming movements was just as long as on the previous occasion. This period, therefore, was considered as representing the time necessary to recover from the anesthesia, and not a period of rapid learning. Later he grew salamanders in isolation, free from the light and vibration. When light was flashed on them after the period when swimming might have been established it was observed that they swam freely, in no wise different from the controls who had been brought up under ordinary laboratory conditions. Despite these extraordinary findings Carmichael (1928) concludes that:

> A knowledge of the developmental mechanics of the nervous system precludes the use of these results as a final confirmation of the theory that the development of behavior is merely a maturation of the native factors. On the contrary, viewed in its largest aspects, the development of behavior seems to be the result of the *interdependent* influence of the action of both environmental and hereditary factors.

In his presidential address Carmichael (1941:17) took occasion to retract some of the implications of his earlier statements:

> When I wrote my first papers in this field. ... I was so under the domination of a universal conditioned reflex theory of the development of adaptive responses that I denied categorically the truth of the statement just made. [The growing animal functions in a way that is in general adaptive at every stage] But every experiment that I have done in the field of the early growth of behavior has forced me to retreat from this environmentalist hypothesis. Now, literally almost noth-

ing seems to me to be left of this hypothesis so far as the very early development of behavior is concerned.

Matthews and Detwiler (1926) also using the chloretone technique to restrict the activity of *Amblystoma* embryos, varied the concentration of solution as well as the duration of immersion. They found that the onset of neuromuscular movement after the embryos were returned to tap water was contingent upon the strength of the chloretone solution and the duration of immersion. Normal reactions were exhibited by embryos immersed only seven or eight days, but an atypical response followed prolonged use of chloretone exposure to anesthesia for more than thirteen days generally resulted in atypical and feeble responses, though the animals continued to live. These findings seem to indicate that restriction of function is effective if the restriction is enforced beyond a critical period. So again we are reminded that the issue is not a simple one of restriction of activity versus practice, but that the stage of development at the inception of the experimental factor (restriction or exercise) and the duration of such factors are of major importance.

Mowrer (1936) who sutured the eyelids of young squabs until they were five or six weeks old, points to a distinction in the developmental processes involved in vestibular nystagmus and optokinetic nystagmus. Vestibular nystagmus in these birds, deprived of normal vision, was both quantitatively and qualitatively indistinguishable from that of birds with normal vision, whereas optokinetic nystagmus was not established until three days after vision had been restored. This rapid development after the restoration of vision he attributed to learning, or, in any event, the maturational factors involved were not alone sufficient to bring these responses to functional maturity. Metfessel (1940) kept roller canaries in soundproof cages so they could not be exposed to the songs of other birds. Selected tones were introduced into some of the cages. Subsequently, all the birds were placed in flight cages where the songs of other birds could be heard. He found that the isolated birds developed rudimentary elements of a species song. The birds exposed to experimental tones showed modification of song elements in accordance with the environmental factor, and the songs of all the birds underwent alteration after they were placed in flight cages.

Dennis (1934a) made a survey of case reports on patients whose sight, either during late childhood or adulthood, had been restored by surgery, for the purpose of culling any information available as to the presence of unlearned behavior as a result of the sudden restoration of vision. Although the results were inconclusive, the author felt that there was no evidence of an unlearned control of behavior by vision. Dennis (1935; 1938) utilized the restriction method in an experimental study of the development of infants. Twin girls were reared from the first until the end of the fourteenth month under nursery conditions which provided the minimum social and motor stimulation. When the development of these infants was from time to time compared with standard norms, it was found that in certain motor achievements they were retarded beyond the upper age period for the appearance of these items in normal groups. The writer was of the opinion that this retardation should be attributed to the restriction of motor practice; on the other hand, he did not find the customary social stimulation indispensable to normal behavioral development.

According to Danzinger and Frankl (1934) extraordinary restrictions are imposed during the first year upon infants reared in Albanian culture. It is reported that the babies, bound to small wooden cradles, are released only for cleaning, not even for nursing. These investigators tried to rate Albanian infants according to the Viennese baby tests. Measured by these standards, the Albanian infants showed retardation in motor development, especially during the third year. In social reactions they showed no retardation. It was in this sphere that they received the greatest amount of stimulation.

Dennis (1940) studied among the Hopi and Navaho Indians the effect of differences in culture upon the development of infants. One striking difference between Indians and Americans is the relative amount of cradle binding or restriction of general motor activity during the first six or twelve months. During the first three months the Hopi child is bound to the cradle all the time except for approximately one hour daily. After that time the periods of freedom increase. Despite these physical restrictions Dennis found that the fundamental development of Indian babies during the first year was not in any appreciable degree different from that of American infants.

The study of maturation has been attacked not only by the restriction of activity beyond the supposed ripening period but also by the introduction of selected environmental factors, particularly increased exercise or practice of a function, into the experimental situation. The overall increase in performance above the norm is usually attributed to the special factor of practice or training. The number of studies bearing upon the effect of specific training is voluminous. In this chapter only a few of those which have particular bearing upon the maturational theory can be mentioned. A common method of study in investigations of this type is known as the equivalent-group method. Groups are equated according to accepted standards—chronological age, mental age, initial achievement, etc.—and then the experimental group is given special exercise in selected activities for a period of time. At the end of the practice periods the achievements of the two groups are again compared, and the superiority (if such is found) of the experimental group over the control group is attributed to the special factor of exercise. Jerslid (1932) used this method in testing practice-effect on mental, motor, and musical performances of over 200 children ranging in age between 2 and 10 years. Practice extended over a period of 6 months. Both groups at the end of that time showed an improvement over their initial tests, but the experimental group showed greater improvement than the control. Innate endowment seemed to be reflected by individual differences in the ability to profit from training. Hilgard (1932) trained an experimental group of young children in such activities as buttoning, cutting with scissors, and climbing a ladder. After 12 weeks of practice the experimental group exceeded the control group on all tests, but one week of practice of the control group at the later age period is sufficient to bring their scores to a level of the experimental group after 12 weeks of practice. The greater profit from training in a shorter period of time by the control group led to the conclusion that other factors than training (partly maturational factors and partly practice in related skills) contributed to the development of these three skills.

Gesell and Thompson (1929) introduced the co-twin method in their famous study of identical twins in stair climbing and cube building behavior. When the twins were 46 months old both infants were at the threshold of stair climbing and cube building behavior. At that time training of twin T was initiated. For 10 minutes each day over a period of six weeks she was stimulated to climb stairs and build a tower with cubes. At the end of this six-week period of training twin T, twin C was started on a 2-week training course in the same activities. Although twin T was more agile in these performances, the investigators were much impressed by the fact that twin C on this first occasion actually climbed the four nursery steps which figured in the experiment, and after 2 weeks of practice twin C was performing these activities as well as her sister had done at the end of 6 weeks of similar training at a slightly earlier time. On the basis of these findings, the authors, claim that: "There is no conclusive evidence that practice and exercise even hasten the actual appearance of types of reactions like climbing and tower building. The time of appearance is fundamentally determined by the ripeness of the neural structures."

It might be noted in this connection that, since the infants were at the threshold of their climbing and tower building behavior at the inception of the training period, the experimental practice had nothing whatever to do with the "actual appearance" or emergence of these types of reactions. It deals rather with the effect of practice on the *expansion* of these activities. This difference is an important one in a discussion of the nature of maturation. Furthermore, when the protocols submitted in the synopsis of the monograph are reviewed, the effect of practice in stair climbing is more evident than the writers' conclusions imply. At the inception of twin C's training period, twin T is admittedly more skillful; 10 weeks later she is still noted as more skillful; 16 weeks later she is reported as more agile, walks faster, and is less afraid of falling; and even as late as 26 weeks after the onset of twin C's training period twin T is reported to be more mobile than twin C, traverses more ground in play, and generally shows more abandon in her motor activities. The issue involved in this study really is not whether practice had any effect upon the emergence of these activities, but rather the relative effect of practice if it is introduced at one time or 6 weeks later. This investigation not only introduced the co-twin technique but also served to stimulate the maturation-versus-learning controversy.

Subsequently these same twins were used as subjects by Strayer (1930) and Hilgard (1933) in studies of different behavior phenomenon. In general, both investigators support the maturational thesis. Influenced by these conclusions of the co-twin studies, especially the inference that motor behavior of infants was not subject to appreciable improvement through practice, but at the same time recognizing that older children and adults do improve in performance through exercise, McGraw (1935) undertook a study with the idea of determining the age when children would begin to show improvement from practice influences in motor performances. From the age of 21 days until he was 22 months old, Johnny, one member of a set of twin boys, was exercised in those motor activities of which he was somewhat capable, whereas his twin, Jimmy, was kept in a crib during the days at the laboratory so that, comparatively, his exercise was restricted. As the children grew, additional behavior items were added to the practice repertory. From time to time the behavior development of these two children was compared to that of a group of children in whom the same activities were being observed. In discussing the concept of increased and restricted exercise upon the development of the various activities considered, it was found convenient to group the activities into two types according to their susceptibility to practice influences. Phylogenetic activities, or those behavior patterns which every child must acquire in order to function biologically as a normal human being (not necessarily as a civilized human being) are more fixed and subject to less modification through mere repetition than are ontogenetic activities, that is, those activities which an individual may or may not acquire.

> The extent to which exercise of an activity may alter the development of a particular behavior course in infancy is contingent upon the following conditions: (1) the neurostructural level at which the activity is controlled; (2) the state of plasticity or fixity of the behavior course at the time increased exercise or use is introduced; (3) the state of fixity attained by the behavior pattern at the time the factor of special exercise is withdrawn; and (4) the phylogenetic origin and importance of the behavior pattern.
>
> Those behavior-patterns which have achieved a high degree of fixity and are controlled at an infracortical level are subject to no appreciable alteration through mere repetition of the activity during the post-natal development of the subject. Also phyletic rudiments of behavior patterns controlled at an infracortical level are resistant to influence or alteration of any significance by increased exercise of the activity. Those phylogenetic activities which succeed

these infracortical rudiments, that is, the kindred activities which are governed at a higher structural level, can be modified in minor details through individual exercise of the function. ... Activities of ontogenetic origin can be greatly accelerated through exercise of the performance, but the degree to which they can be modified is dependent upon the state of maturation or plasticity of the behavior pattern at the time the factor of exercise is introduced (McGraw 1935:309).

The writer emphasizes that there are critical periods when any given activity is most susceptible to modification through repetition of performance. Repeated observations (McGraw 1939b) on these two boys over a period of years have revealed that the child given greater opportunity for motor activity during the first two years still exhibits superior motor coordination.

McGraw (1940b) also used the co-twin technique in studying the effect of training in the achievement of voluntary micturition. Two sets of identical twin boys were used as subjects. After the first few weeks of life one member of each pair was placed on the chamber for voiding at hourly intervals during seven hours of the day until they were 17 months of age respectively. Their brothers were not permitted to use the vessels until they were 14 months of age in one instance and 24 months in the other. From the achievement curves of their four children it can be seen that the experimental subjects did not profit by the long and systematic training program. It was pointed out that during the early months the reflex mechanism controlling micturition was so hypersensitive that any mild handling or disturbance of position might initiate voiding. The slight elevation in the achievement curve at this time was attributed to the hypersensitivity of the reflex mechanism. As the reflex became less sensitive to vicarious stimuli, a decline in the achievement curve was noted. Later a rapid rise in the percentage of successful response to the vessel paralleled the onset of cortical participation in the act of micturition, the cortical participation being reflected in the child's behavior by unmistakable signs of awareness of the act and the result. As the cortical influence shifts from simple association to more complex discrimination and generalization a second decline in the achievement curve is evident. The functional integration of complex cortical processes is reflected in a second rapid rise of the achievement curve, after which the performance is stabilized at a fairly high level.

It is evident from the discussions of these studies of practice effects upon behavior development that there are two fundamental problems involved in the maturational theory. The first concerns the genetic components of behavior development and its corollary, the modifiability of behavior through environmental influences; and the second concerns the relationship between developmental changes of the nervous system and behavior as it is manifested during ontogeny.

Determination of Neural Status
and Associated Behavior

It is safe to say that those who think of maturation as a term for describing certain changes taking place within the nervous system before function can be detected do not presuppose that these changes are in the nature of chromosonal rearrangements. In this application, the term refers generally to some other type of anatomical or chemical change which develops without reference to neuromuscular functioning. And the real issue of the maturational theory is whether or not the neurostructural changes are prefunctional or whether the structural changes are molded by somatic function out of a rather homogeneous neural mass. That is, does function determine the neural organization or does the neural organization form the framework within which function takes place.

Investigations bearing upon these problems are in general of the following order: (1) studies in neurogenesis which reflect in a general way the developmental character of embryonic nerve tissues; (2) observation upon the development of overt behavior; and (3) determination of a relationship between specific somatic activities and certain anatomical and physiological constructs in a time-space framework.

In embryological studies of the ontogeny of the nervous system there are some impressive observations the interpretations of which are basic to an understanding of neuromuscular functioning. In the first place, developmental changes in neural morphology—subdivisions, bulges, folds, fissures, etc.—are strikingly constant in all normal individuals of a species. Not only is it well substantiated that nerve cells of a given structure, location, and organization demonstrate functional peculiarities, but, furthermore, it is possible to detect certain morphological and histological differentiation in an early embryonic stage. It is possible to know as soon as the medullary plate is laid down those divisions that will develop into brain and those that will ultimately develop into spinal cord. The question of intrinsic and extrinsic factors is aroused as soon as the medullary plate assumes more definite form. Development at this stage is indicated by an increase in thickness and mechanical rigidity, attributed to an increased height of the cells. Embryologists ask whether the cells possess some intrinsic power to grow in height or whether they are compressed into elongation by the surrounding ectoderm. Similarly, does the medullary plate possess an independent power to form a neural tube, or is it dependent upon the adjacent ectoderm? Experiments in this connection lead to the conclusion that "the medullary plate and non-neural ectoderm are both taking an active, cooperative, part in the movements of neuralation. Although capable of manifesting their individual abilities independently they normally do team work by assisting and reinforcing each other" (Weiss 1939). It is evident that the problem of neurostructural development and its relations to the surrounding environment is one that arises early in embryology. It is also evident from the citations of (Weiss 1939) that conflicting results may be obtained.

Many of the transplantation and extra-plantation experiments of embryonic neural systems provoke thought concerning the growth qualities of nerve tissues. There is considerable evidence too that the problem is not simply a matter of determining a relationship between neural cells and their environment. The element of time is highly significant. The exciting experiments of Spemann (1925) brought out the point that there is a critical period when the presumptive neural plate becomes determined and will continue to develop into neural tissue despite its transplanted location. Transplantation at an earlier stage, however, showed differentiation in accordance with the new environment.

Detwiler (1920; 1928; 1931) showed that during later stages of embryonic development of the *Amblystoma* a transplant of a limb bud caudad would attract the brachial nerves away from their normal pathways in order to innervate the displaced limb bud. It was also shown that the force of attraction which a transplanted limb may exert upon the brachial nerves operates only through a relatively short period of time. Furthermore, the influence appeared to be nonspecific, as transplants of nasal placode or eye grafts would exert an attractive force upon spinal nerves with which they normally had no connection. Apparently it is the state of rapid differentiation which influences the direction of spinal innervation. It has been suggested that an electrical field may be set up as a result of the high physiological activity at the focus of rapid differentiation.

Gilchrist (1933) also points out that there is a crucial stage in the development of amphibian embryos when the neural fold is responsive to thermal changes. He found that if eggs were warmed on the right side during the blastular stage they developed a larger neural fold, whereas a comparable increase of the neural fold was not educed by warming during the period of early gastrulation. He points out that the period of increased susceptibility to thermal stimulation corresponds to the time at which the determination of neural plate material was most actively under way.

Durken (1932) has shown that at the end of the process of gastrulation of Triton embryos presumptive epidermis and medullary material are no longer interchangeable. If they are interchanged after this stage, they develop not in accordance with their position but in accordance with their origin. These studies of the embryonic nervous system bring up several considerations which are pertinent to an understanding of the process of maturation and development. They suggest that presumptive neural tissue is first in an indeterminate state when it is more or less susceptible to particular environmental forces. They suggest further that there are critical periods in the development of any phenomenon when it is most responsive to extensive stimulation, and there is some indication that these crucial periods correspond to periods of rapid differentiation and development.

Other writers have felt that the type or quality of early neurosomatic movements is of special significance to the maturation theory, and that the relationships between neural structures and functions are reflected in the quality of these embryonic movements. The central issue concerning these early fetal movements seems to revolve around a controversy as to their localization or specificity in contradistinction to a total body or mass response. The controversy has reference to theories of development as a process. If the earliest detectable fetal movements are in the nature of localized specific reflex activities, development presumably proceeds from the simple to the complex. On the other hand, if the initial neurogenic movement is of the nature of mass or "total body" response, the process of early development is one wherein the local or discrete response gains specificity out of the general mass matrix.

Such controversies on the nature of the developmental process have served to stimulate intensive observations on the primary embryonic and fetal movements of many different species. Some have based their interpretations entirely upon the nature of behavior response without essaying to determine the developmental state of neural counterparts. Bridgman and Carmichael (1935) who worked with fetal guinea pigs, report that the order of neuromuscular reactions is (1) myogenic, (2) local response to external stimuli, and (3) spontaneous activities. Tuge (1937), utilizing the saline solution technique in studying embryonic movements of carrier pigeons, also reports myogenic prior to neurogenic movements. The earliest neuromuscular action appears to be a lateral flexion of the neck, a reaction which extends until it embraces the entire axial musculature. Kuo (1939), from his observations of chick embryos, says that so-called local reflexes may appear during any stage of development, though there is a steady increase in the percentage of local responses as the fetus develops. Tracy (1926) also reports discrete muscular activity as primary in the toadfish. According to Swenson (1929) the first movements of the fetal rat consist of a lateral flexion of the head, trunk, and rump; and essentially the same type of movement is reported by Pankratz (1931) as characteristic of the fetal rabbit. The same type of unilateral head and upper-trunk flexion is reported by Coronios (1933) as the primary neuromuscular movement of the fetal cat. Indeed, one is again and again impressed when descriptions of embryonic movements of different species—birds, fishes, and mammals—are presented that the unilateral flexion of the cervical axis is such a common occurrence.

Other investigators have supported their observations of overt behavior by cytological analysis of the nervous system of the experimental fetuses. The now classical researches of Coghill (1929b) are demonstrative of the knowledge which may be gained by such colossal and tedious methods.

Coghill has not only determined the neural status of the nervous system in Amblystoma embryos at the onset of neurogenic behavior but also has followed it through all the progressive stages to adulthood. Angulo y Gonzales (1930; 1935) has applied essentially the same methods in studying structural and functional correlation as evidenced in the fetal rat. According to Tuge (1931), early embryonic movements of the turtle are essentially of the same nature as those of other embryonic vertebrates although in adulthood the vertebral axis of this animal is attached to his shell. Windle and his collaborators (1930; 1932; 1933; 1934), in their studies of fetal chicks and fetal cats, have been concerned with the nature of the first neurogenic movements of embryos as well as with the structural counterparts of fetal behavior. Tilney, giving particular attention to the structural maturation of the rat brain, has shown parallel changes in behavior and cerebral maturation not only during the fetal period but also during postnatal development until the rat achieves adulthood. Tilney (1933) interprets his findings in the light of phylogenetic as well as ontogenetic development. There have been a few studies of the primary neuromuscular movements of the human fetus. The early reports (Strassmann 1903; Yanase 1907; Krabbe 1912; Minkowski 1922; Bolaffio and Artom 1924) antedate the current interests in the quality of primary behavior responses, though Coghill has interpreted Minkowski's observations in terms of his own theory of "individuation." Current work on the human fetus will be discussed more fully later. Since the discussion of fetal behavior will be more amply reviewed in another chapter of this book, only those implications concerning the nature of maturation and development will be mentioned here.

As indicated above, the writers on the subject of primary embryonic movements seem to be divided into two schools of thought. Since the classical challenge of Coghill, there has been one group, including Coghill, Angulo y Gonzales, Tuge, and, more recently, Hooker (1936; 1937), who assert that neurogenic behavior in its origin is in the nature of a total pattern which, from the beginning, is integrated; local reflexes, or independent limb movements, are "individuated" out of this dominant pattern. Windle, Carmichael, Tracy, and Kuo claim that local reflexes may be primary, or at least that local reflexes occur as early as the total body response. The issue is not merely an academic one but one which is basic to an understanding of the process of development. According to Coghill (1933):

> An action is regarded as total when it involves all the muscles of a functional system that are capable of responding at the time. ... Action appears first in the anterior part of the axial musculature, and spreads thence tailward through the axial system, and then into the appendicular system, so that before an appendage can act on its own it acts only as an integral part of a whole, which is axial and appendicular.

Individuation is the process whereby appendages or local muscular groups achieve motility somewhat independent of the axial system. He points out that individuation is not the same as specialization or specification, which involve an adjustment of the organism to the environment, but it represents rather "a definite and peculiar relation of a part of the organism to the organism itself as a whole." (Coghill apparently uses the term "integration" as referring to the unification of movement, although groups of muscles are involved. Other authors, including the writer, apply the term to the smoothness and articulation of various movements which constitute an activity. In this sense, particular muscles may be capable of independent action but they function as an integral part of the larger activity under consideration, whereas, in Coghill's analysis of early embryonic movement, local muscle groups are incapable of independent action). Windle and Orr (1934) draw a distinction between responses to mechanical stimulation and spontaneous motility. They state that the first response of the chick embryo to mechanical stimulation is a local reflex but that at the same time spontaneous motility in the form of well-coordinated swimming movements is present. Coghill (1940) has recently summarized the principal contributions of various investigators concerning the nature of primary embryonic movements.

The descriptive protocols of the embryonic behavior of various species by different observers show striking consistency in the manner in which certain types of embryonic movements become organized. Movements begin by flexion in the cervical region, gradually expand until the trunk is involved, then a contralateral movement is initiated in the cervical region. The two movements become synthesized and finally consummate in a swimming movement. Subsequently, the movements of the appendages achieve independent action. These fundamental movements are basic to the development of progression. In addition to the priority of spontaneous or stimulated reactions, the issue seems to be whether mechanical or electrical stimulation activates the basic pattern or whether such stimulation elicits a local response. Basically, the question at stake is whether reactions to exogenous stimulation develop in a manner different from the organic, phyletic neuromuscular activities.

These investigations of embryonic and fetal behavior (particularly those of Coghill and Windle and their collaborators) are of further significance because in some instances they represent monumental efforts toward the determination of the relationship between overt behavior development and neural structures. Coghill (1933) has emphasized the importance of determining the neural structures correlative with the observed behavior but adds:

> Argument as to whether structure precedes function or function precedes structure, or in other terms, whether structure causes function or function causes structure is beside the mark, for neither can exist without the other at any point

in space or time; they merge in fact, into a space-time relation. ... If we think of structure *in* function as a space-time relation, as we must in a purely scientific discipline, we may hope to attain to [sic] all the understanding of behavior that science has to offer. The embryological method ... has transformed anatomy into a science, in so far as it has correlated time relations with space relations which constituted the anatomy of other days. Also, it is now engaged in a similar transformation of physiology through the contribution of such concepts as totipotence, pleuripotence, organizers, gradients, all of which have meaning only as the organism is regarded, not as a static pattern in space, but as a dynamic pattern in time.

It is fitting that these new concepts of organizer, gradients, electrodynamic fields, etc., should be given at least passing consideration, especially as they have reference to an understanding of the meaning of maturation of behavior. According to the old theories of heredity there was, strictly speaking, no problem of development distinct from the problem of heredity. Modern theories of heredity contend that every cell inherits a complete set of chromosomes from the germ plasm, but they do not explain how, during the course of ontogenetic development, cells differentiate and become distinct, both structurally and functionally. At the present time the solution of this problem has scarcely advanced beyond the stage of postulates. The most we can hope for is an evaluation of the various hypotheses, an interpretation of experimental results in terms of the most acceptable logical system. Spemann (1925) has applied the term organizer in reference to embryonic cells capable of inducing the formation of new Anlagen; other writers have made more general application of the concept as referring to chemical constituents which operate in effecting cellular and organ differentiation. Child (1924) proposes the term physiological gradients, with the explanation that the primary factors in determining organization and differentiation are regional differences in the rates of metabolism. Furthermore, these differences in rates of metabolism are organized along an axis pattern in a gradient system and centers of high metabolism are centers of dominance, influencing and determining the rate of more remote regions. In discussing the concept, Child (1939) says:

> The gradient conception in its relation to development is a working hypothesis which although based on many lines of evidence is, like other conceptions of development, subject to modification as experiment progresses. It is essentially a conception of the living organism as primarily not a mosaic of chemical substances, but a dynamic system in which the fundamental activities of the species protoplasm concerned are the ordering, determining and integrating factors. ... Moreover, the primary dynamic system is regarded, not as autonomous, but as directly or indirectly a product of environmental factors as well as of heredity; in short, as a behavior pattern of a protoplasm of specific constitution.

Burr and Hovland (1937) and Northrop and Burr (1937) have formulated a theory of biological organization which is based primarily upon the bio-electric properties of living protoplasm.

> The fundamental thesis of this theory is that physical, philosophical, and bio-logical considerations warrant the extension to biology of the hypothesis that 'The pattern of organization of any biological system is established by a com-plex electrodynamic field, which is in part determined by its atomic physico-chemical components and which in part determines the behavior and orienta-tion of those components.' This field is electrical in the physical sense (Northrop and Burr 1937).

Weiss (1939), in discussing the field concept, admits that it is an abstrac-tion but since practically all developmental phenomena exhibit field-like characters there is reason to believe that it is an expression of physical real-ity. This is scarcely the place to embark upon a discussion of the shades of differences in the concept of biological fields as presented by various experi-menters, or of the polemics which have arisen over the merits of the various concepts, gradients, fields, organizers, etc. Weiss (1939) and Child (1940) have each ably criticized the points of view of the other. These references serve here merely to indicate that the conceptual frameworks for interpreting biological phenomena are becoming progressively dynamic, and that the concepts of biological differentiation are pertinent to an interpretation of data concerning complex human behavior. In the last chapters of his book Child (1924) has called attention to the parallel between the simpler biologi-cal processes and social integration.

In the meantime the study of ontogenetic development is aided by the in-troduction of more dynamic concepts into modern theories of genetics. Cur-rent ideas as to the role of genes in ontogenesis have been lucidly appraised by both Stern (1939) and Waddington (1939). It has been pointed out that no single formula, such as the chromosome theory, is adequate to envisage all recent observations on the connection between the gene and the individ-ual character with which it is associated. Obviously the gene is no longer accredited the specificity of its early days. It is now recognized that specific characters (phenotypes) are determined by the interaction of many genes, and that any one gene may affect many characters. There have been many studies (Stadler 1939; Dunn 1940) which show the genes themselves and their chromosomal arrangement to be subject to alteration through external agents, such as the roentgen ray. Furthermore, it has been demonstrated (Danzinger and Frankl 1934; Dunn 1940) that the introduction of certain environmental factors (temperature changes) at critical periods can produce changes in a developmental process in the same way that gene changes or other mutations do. Whatever the gene produces is contingent upon the gene combinations and the conditions under which the organism develops. Goldschmidt (1938) has reported studies which seem to indicate that various characteristics are the result of changes in growth rates and that the genes exercise a controlling influence upon the rates of development. In studying the family of Cucurbitaceae, Sinnott (1937b; 1939) has observed that the relationship between width and length remained constant despite the great variety of forms which they finally display. Sinnott (1937a) thereupon concludes:

> What is inherited, and therefore what genes control, seem to be these constant growth relationships. As growth proceeds, the proportions of parts change, complexity increases, and the familiar developmental story unfolds. Running through all this complexity, however, is a basic constancy, the inherited growth relationship, established from the beginning. It should be possible to determine for any organic pattern a series of constants of this sort. If genes control relationships between rates of growth in various dimensions, they may perhaps control relationships between rates of other developmental process which are not spatially arranged, notably the complex series of chemical changes concerned in the development of many traits. This is opposed to the view that the gene initiates only the first step in such a series, the later ones bearing no resemblance to the original genic impetus.

At the present time one of the most promising methods of approach in attacking problems of development seems to be the analysis of the interrelation of the rates of growth manifested by the various aspects of development. Huxley (1931) has shown that although different parts of an organism may grow at different rates the ratio between them is a constant and Sinnott and Dunn (1939) as well as Goldschmidt (1938) seem to regard growth rates as the primary domain of the gene. These suggestions are especially promising, as the determination of growth constants may provide a common denominator whereby evidence assembled in the various disciplines may be comparably interpreted. It is conceivable that determinations of the relative growth rates of various aspects of behavior development may provide a technique for comparing activities which are predominantly maturational in their process of development with those which are the result of training, or distinctly ontogenetic in origin.

These speculations, however, are far ahead of the game. At the present time ontogenetic studies of both architectonics of the central nervous system and overt behavior of the human being are still in an exploratory stage. Before much can be accomplished toward determining the correlation of structure of nerve tissues and overt behavior additional descriptions of both aspects of behavior must be accumulated, and accumulated in such a manner that they can be subjected to symbolic manipulation.

Neural Maturation of the Central Nervous System of the Human Being

Infant

The paucity of data on the structural maturation of the human nervous system is understandable. In the first place, it is impossible beyond the fetal period to obtain functional or behavior data and analysis of the neural structures of the same subjects. Since human beings are not born in litters it is impossible to secure comparable data on different subjects, as has been done with lower animals. Moreover, the technical labor involved in analyzing one brain or one segment of the nervous system is colossal. Despite the promise of modern electrophysiological techniques, the best evidence of neural maturation is still obtained from histological sections. The methods employed in such studies are well known, though the criteria used as evidence of structural maturation vary and some of these criteria have been the subject of controversy.

One of the earliest reports of structural studies of the human cortex is that of Bolton and Moyes (1912) who made a cytological analysis of the cortex of a human fetus of eighteen weeks. After being hardened in formalin, the sections of the brain were stained both by the polychrome and iron-alum-haematoxylin methods. It was found that at this stage there was sufficient cellular differentiation to distinguish five local areas. An area of pyramidal cells (Betz cells) as mapped out in the frontal parietal region. Even at this early stage the area of Betz cells, which corresponds roughly to the precentral motor area of the mature brain, shows the most advanced development. The presence of Betz cells and well-defined cortical areas at a time when the fissure of Rolando and the sulcus cinguli are barely perceptible led to the conclusion that histological differentiation precedes fissuration.

Langworthy (1930, 1933) used the Weigert-Pal method of staining postmortem sections to bring out the medullation of nerve tracts in a study of the brain stem of fetuses between 6 months of age, and birth. One 2-month brain was included in the series. Langworthy notes that in general tracts become medullated in the order of their phylogenetic development, but reflex activity may be elicited prior to myelinization. A rapid increase in myelinization occurs in the brain stem between birth and the second postnatal month. Langworthy is of the opinion that the inception of function advances the medullary process. Even in the seventh month fetus the pathways concerned with the fundamental processes of living are well medullated. From these studies of myelinated pathways it seems clearly evident that the infracortical centers are the ones prepared to govern neuromuscular activities of the maturing fetus and the newborn infant.

Realizing that postnatal maturation of the brain cannot be attributed to multiplication or subdivision of the cells, DeCrinis (1932) used the Golgi method of staining so as to determine advancement in dendritic processes, and he considered cell proliferation as a fundamental criterion of structural maturation. He reports analysis of 68 human brains, varying in age from 5 days to 13 years. In the 5-day-old infant's brain preparations from the motor region did not reveal dendritic processes, but in preparations from the brain of a 10-week-old child such processes were clearly discernible. Sections from the motor region of an 11-month-old infant's brain give a picture of advanced development. Pyramidal cells have become strikingly long, the processes are well matured, and the dendrites are clearly impregnated. The onset of rapid maturation in Broca's area occurs later. Although development in this region is from the beginning behind that of the motor area, during the first year this disparity becomes even greater. At 14 months cellular maturation in Broca's area has not attained the stage of maturation reached by the motor region at 11 months, and it does not achieve a comparable stage of development until about the age of 17 postnatal months. Other frontal lobe areas lag behind even more. Although some evidence of maturation in the frontal region occurs at about 11 months, it is not until approximately the fourth year that the apical processes of development mature to a stage comparable to those of the motor area at 11 months. De Crinis further points out that the area of the muscle sense (sensiblen) and the sensory projection areas are the first to mature. He postulates that these early-maturing centers are the cornerstones of intellectual development.

For some time Tilney was interested not only in the ontogenetic relationship between the structure of the nervous system and behavior but also in the phylogenetic evolvement of the brain. To this end he launched an extensive investigation to delineate the structural and functional maturation in each different species (opossum, rat, guinea pig, pig, cat, and man) (Tilney and Kubie 1931). The basic assumption in investigations was that a structural adequacy in the nervous system is essential before specialized reactions can be elicited. The neocortex is a common heritage of all mammals, including man. For this reason Tilney focused much of his attention upon the development of this organ. Three methods were employed in the analysis of structural maturation: (1) organogenetic studies by means of the Born method of reconstruction, which discloses the chief features in the organic differentiation of the brain, (2) histogenic studies which reveal the maturing processes of nerve cells and their fibers; and (3) the myelinogenetic studies which indicate the final ripening processes of the nerve cells. Much of his work on the structural maturation of the brain of lower animals has been published, and in many of these reports references are made to the basic maturational processes of the human cortex. Structural studies, at monthly intervals, of the human brain had been completed for the fetal period, including the brain of the infant at birth.

In unpublished reports of this phase of his investigation Tilney has found it convenient, both on the basis of structural differentiation and qualities of behavior, to draw a distinction between two major types of behavior, namely, nuclear activities and cortical behavior. On the structural side nuclear regions are recognized by the way in which the cells are clustered together without much distinction or order in arrangement and also by cytological distinctions in the size, shape, and dendritic processes of the nerve cells. The nuclear regions are much older phylogenetically than the cortical areas, in fact as ancient as the beginning of vertebrate life. During ontogeny they mature earlier than the cortical areas do. Behavior controlled at a nuclear level exhibits a sudden turnover of impulses without much latency in the period of delivery. Structurally, the cortex shows an orderly arrangement of cells, and cortical behavior discloses a planned element in the reaction, a latency and variety of response.

According to Tilney (1937) there are three major phases in the development of the cerebral cortex:

(1) General cortical differentiation is characterized by a series of migratory laminations, resulting in a six-layered neocortex.

(2) Divisional cortical differentiation denotes the time when four major cortical divisions can be delineated, namely, the bulbar cortex, the paleocortex, the archicortex and the neocortex. Of them the first to appear is the archicortex which, Tilney believes governs vital functions having to do with such behavior activities as hunger and thirst, elimination, sex behavior, and such cortical activity as may be involved in breathing and pulsation the heart.

(3) Local cortical differentiation, the third major phase, is distinguished by the rearrangement and disposition of cells which produce such local areas at the motor, auditory, sensory, and visual, recognized in the mature brain. During this period there is marked alteration in cellular structures. There is a tendency for the cells to become pyriform or pyramidal in outline and to show a relative reduction in granular elements.

General cortical differentiation is indicated in the brain of the 6-week-old embryo; by 7 weeks there is the beginning of primary migratory lamination; by 9 weeks primary migratory lamination has become well defined; 10 weeks marks the onset of secondary migratory lamination; and the 13-week-old embryo shows the beginning of tertiary migratory lamination. This process of lamination continues until in the 4-month fetus the six cortical layers are laid down, though at this time they are probably contributing nothing to behavior. Not until the seventh month is the migratory process completed. Fissuration can be detected earlier, and, although the 6-month fetal cortex is almost homogeneous, it is approaching the inception of local differentiation. At 8 months increased development in depth and richness of fissuration is indicated. Although there are incipient signs of local cortical differentiation during the last stages of fetal development, even in the cortex of the newborn infant there is not sufficient local cortical differentiation to indicate particular areas are concerned with function. Apparently the cells are all there in readiness for final ripening.

About the same time that Tilney extended his structural functional investigations to include the human, Conel (1939a; 1939b) independently began a program of studies on the development of cytological structures of the human brain, using the cortex of the newborn infant as the point of departure. Conel utilized several staining methods in order to educe developmental changes in cell bodies and fibers. Structural features which were accepted as criteria in estimating advancing development were: (1) size of cell body, presence of neurofibrils, size and length of cell processes, (2) number, size, and length of horizontal and vertical fibers in the cortex, (3) the density of the neuropil in the various horizontal laminae, and (4) the quantity of myelin present. So far Conel has published reports on the newborn brain only, though analyses of more advanced stages are under way. Even in the newborn cortex the cells have achieved their adult arrangement, which is in agreement with the observations of Tilney that the cellular division and migratory processes are complete at birth. According to all criteria, the cells in the most advanced stage of development at the time of birth are those in the anterior-central gyrus. More specifically, that part of the anterior central gyrus which mediates movements of the neck and shoulders is more advanced than any other part of this area. The most mature area is localized in the anterior half of each of the three frontal gyri.

Conel has recently been able to extend his publications on the development of the cerebral cortex to include that of the one month infant. He has noted the difference between the newborn and one-month brain and points out that the greatest change has taken place in the region of the hand of the motor area of the gyrus centralis. In all functional areas, however, the development of the cortex during the first month is but slight, and whether the neurons of the cortex are functioning with respect to conducting nerve impulses is still questionable.

Behavior Development of the Infant

The abundance of evidence gleaned from the several studies of the structural maturation of the newborn infant's brain, renders functional participation of the cortex problematical at this stage of development. Recently Hooker (1939) has been conducting extensive investigations of fetal behavior and has obtained data on 41 specimens ranging in age from 7 to 28 weeks of menstrual age. Hooker reports that he has elicited movements in response to tactile stimulation shortly after 8 weeks of age. At this age response is elicited only when the area supplied by the mandibular and maxillary divisions of the fifth cranial nerve is stimulated. Although the nature of the response was somewhat diffuse involving all neuromuscular mechanisms capable of function at the time, the response to external or mechanical stimulation apparently preceded overt response to endogenous stimulation. Spontaneous movements were first observed in a fetus of 9 weeks of menstrual age. By the end of 14 weeks most of the neonatal reflexes could be elicited. Hooker (1937) has presented an evaluation of Minkowski's findings in light of his own work. He believes that both Minkowski's and his observations support the Coghillian theory of individual as to the process whereby local reflexes emerge. Determination of the neurostructural maturation of the nervous system of the fetuses in Hooker's investigation has not been reported, though the plan of investigation embraces this aspect of the behavior as well as function. Endocrine status and electrocardiograms are being made upon the same specimen. For a more complete summary of the various reports upon fetal behavior, the reader is referred to Munn (1938).

In many respects the neonatal period is the one most available to investigators prior to school age. There have been many studies apprising the behavior repertoire of the newborn infant. An extensive bibliography of such investigations was presented by Dennis (1934b) as well as a classification of responses reported. Selected bibliographies on behavior at this stage of development have also been assembled by Pratt (1930) and Dewey (1935). In general, most of these studies are in the nature of inventories listing the types of reactions with which the newborn infant is equipped.

During the past quarter of a century volumes have been accumulated on the behavior development of the infant and growing child. Much of the earlier work consisted of biographical reports of the achievements of individual children, of which Preyer (1888), Shinn (1900), and Fenton (1925) are familiar references.

The general arousal of interest in the behavior of the infant and preschool child found expression in the psychological and educational laboratories throughout the country. The prevailing thought operating in most of the early investigations was that individual differences in the development of children are so great that only norms or averages of achievement, based upon the study of large groups of children within given chronological periods, would be of scientific value. The period saw the production of a number of standardized tests or scales for the purpose of measuring developmental achievements. Gesell (1925) was a pioneer in applying this technique to the study of infant development. Application and modifications of the Gesell tests have been presented by other investigators, notably Buhler (1930) and Bayley (1939). Stutsman (1931) and the Minnesota group (Goodenough, Maurer, and Van Wagenen 1940) published standardized scales for the measurement of behavior development during the preschool period. Most of these tests consisted of listing in order selected items of behavior, the order being determined by the chronological age at which the "average" child would presumably achieve that particular behavior. There was no reason to assume that the achievement of items listed in one chronological period was directly related to the achievement of behavior items listed in a subsequent chronological interval. But standardized measures provide information as to when particular behavior items are achieved they do not disclose how function undergoes change as it progresses from inception toward optimum efficiency. It is also recognized that since children do not grow uniformly the average norms do not reflect the actual course of growth as represented by any one individual. A child may be advanced in one aspect of development and retarded in another.

Recognition of this aspect of normative scales aroused interest in the longitudinal, as distinct from the cross-sectional, methods of collecting data on behavior development. Shirley's study (1931a; 1933a; 1933b) of twenty-five infants over a period of two years reflects both these methodological influences. Although she presents descriptive reports of behavior changes in individual infants, the actual treatment of the data is the same as that customarily utilized in purely cross-sectional studies. A further refinement of the longitudinal method was introduced with the realization that in order to analyze behavior development as a process of change not only the child subjects but sequential change in specific behavior phenomena would have to become the subject matter of investigation. The method of analysis calls for an appraisal of behavior, not as it deviates from some reputed average but as it compares to its own functioning at an earlier time. In other words, the behavior phenomenon becomes the subject of inquiry and the reference base is determined by the points of origin and termination. Focusing inquiry upon definite phenomena rather than inventorying mass achievements of the individual is both reasonable and practical; and, logically, the experimental problem is not different from the study of the individual.

> Socially speaking, an individual represents merely nodal points of activity within a Society. ... In much the same way that arbitrary limits can be prescribed around nodal points within a society to signify a particular individual, so may arbitrary limits, for experimental and observational reasons, be set around nodal points of a psycho-motor activity (McGraw 1940a).

Halverson's (1931; 1932) analysis of progressive changes exhibited by the child in prehending various objects is illustrative of the method. Applying this method intensively over a period of years, McGraw (1942) has been able to delineate significant features in the neuromuscular development of a series of activities common to the growing infant. In general the technique employed and the objectives in all these studies have been essentially the same.

In the main the studies focus upon development of the motor activities of the infant during the first two years of life. The investigations were launched under the altogether reasonable assumption that changes in the nature or qualities of muscular movements reflect maturation of the central nervous system. Inventories of the behavior repertoire of the newborn infant indicate that certain types of somatic activity attain a mature stage of development during the fetal period. A newborn infant coughs, sneezes, yawns, etc., about as efficiently as the adult, and qualitatively there is little difference in the somatic aspect of these performances when expressed by individuals of differing degrees of maturation. Although it is quite true that the motor aspect of such reflexes is mature at the time of birth, it is equally obvious that even these typically reflex functions gain a degree of voluntary or deliberate control as the individual matures. A cough or yawn may become organized as a part of one's social behavior, but the motor aspect of that cough or yawn remains essentially the same. For example, at a prolonged social gathering a husband, catching the eye of his wife far across the room, may yawn to indicate he would like to leave, but after the initial deliberate start the yawn is a yawn so far as the motor performance goes (McGraw 1939a). This simple instance of a deliberate quality being superimposed or incorporated into a reflex mechanism illustrates the problem of behavior development as exhibited in the neuromuscular maturation of the growing infant.

McGraw observed and recorded descriptively those changes in motor performances of infants which lead to the consummation of behaviors such as creeping, erect locomotion, the assumption of erect and sitting postures, swimming, adjustment to an inverted position, suspension-grasp behavior, and responses to sudden startle stimuli. In appraising the data, those qualities of each behavior function were selected which reflected the course or trend of development without regard to individual differences or peculiarities. Although the data were recorded originally without interpretative rationale, it became evident in the course of investigation that certain qualities of movement signify grossly the level of neural organization involved. Ratings of the original observations were made with these thoughts in mind. In interpreting the data, it seemed reasonable especially in the light of cytological evidence cited above, to embark with the assumption that the cerebral cortex is not functioning appreciably at the time of birth. Without being specific or attempting to allocate a given function to a particular neural counterpart, it was found both feasible and practical to think of behavior as of two major centers of control, namely, (1) cortical and (2) subcortical or nuclear. It was also recognized that the cerebral cortex is twofold in function. It not only exercises an activating or governing influence upon neuromotor behavior but also exerts an inhibitory influence upon behavior controlled at an infracortical level. In evaluating the observations on changing behavior patterns an attempt was made to point out those qualities which indicate (1) when an activity is under infracortical dominance, (2) when inhibitory influences from the cortex become apparent, (3) when cortical participation in muscular movements is involved, and (4) when the activity attains a comparatively mature state of cortical functioning.

The early reflex patterns of behavior, such as the Moro or body startle and the suspension grasp reflex, are of subcortical dominance. The diminution of such behaviors reflects the development of cortical inhibition. When the startle reaction is followed through its entire course, an increase in the manifestation of the subcortical activity is observed during the first three or four weeks (McGraw 1937). This period is followed by a period of decline, representing cortical inhibitory influences, until finally the characteristic response is merely a body jerk. This is an activity in which the subcortical centers continue to play an active role even during adulthood. After ultimate development is achieved, however, the cortex is definitely engaged practically at the instant of the startling stimulation, and cortical inquiry as to the meaning of the startling stimulus serves to restrict the nuclear aspect of the reaction. It has been asserted that the reflex is of phyletic origin, reminiscent of the "clasping reflex" of primates. The suspension grasp reflex also shows a decline after the first few weeks of life (McGraw 1940c). The period of cortical inhibition extends over several weeks or months and the onset of voluntary suspension is evident when the child exercises his own choice in the matter of suspension.

Many of the typically newborn behavior patterns appear to be of phyletic origin. It has been demonstrated in a study of aquatic behavior that when newborn infants (as early as eleven days) are submerged in water they engage in rhythmical swimming movements simulating those of other vertebrates (McGraw 1939c). The presence of swimming movements in the newborn of many species, including the human, suggests functional evidence of the phylogeny of man. Swimming is probably one of the oldest phylogenetic activities of which there is a residual in infant behavior. Swimming movements of the newborn are controlled at an infracortical level. The onset of an inhibitory influence from the cortex is expressed by disintegration of the swimming movements and the substitution of disorganized struggling activities. Basically, the neuromuscular mechanisms which mediate the reflex swimming movements may be essentially the same as those activated in the reflex crawling and stepping movements of the infant.

In following developmental changes of behavior patterns which lead to deliberate progression it is possible to determine grossly those periods when the movements are predominately nuclear, when the cortex begins to suppress or inhibit the nuclear activities, and when the cortex takes a dominant role in activating or controlling the muscular movements involved (McGraw 1941a). The course of development as manifested in these various activities seems to indicate that subcortical movements show their maximum manifestation at about the end of the first month, and between the first and the fourth month, approximately, there occurs a progressive decline in subcortical movements. The onset of cortical inhibitory influences is observed in the muscles of the upper part of the body earlier than in the region of the pelvis and lower extremities. Likewise, the development of cortical control over neuromuscular activity proceeds essentially in a cephalocaudal trend. One who has observed the development of prone progression in infants cannot fail to be struck by the similarity in the infant's development of progression and the progression of the salamander, as described by Coghill (1929b). At the inception of cortical participation in a given activity the movements are ordinarily staccato and poorly coordinated. Further development is reflected more by increasing integration of the movements involved than by any actual changes in motor pattern.

Unlike the activities of progression, there is no reflex sitting posture in the behavior repertoire of the newborn infant (McGraw 1941b). This activity is of recent origin and probably was never organized at a nuclear level. In most infants cortical control over the function precedes complete cortical participation in the progressive activities. Maturation of the central nervous system is also indicated in the postural adjustments of the infant to an inverted position (McGraw 1940d). The situation has the advantage of being simple, and, at the same time, it is not one which the child would experience through "spontaneous" exercise. The predominant adjustment of the newborn infant is one of general flexion. The flexor movements are under subcortical control. After the first few months the development of extensor muscles becomes evident, as is shown in the arched spine during inversion. Cortical participation in the adjustment is indicated when the child makes a deliberate effort to right himself, and complete integration of the various centers stimulated is evident when the child's understanding of the situation is such that he displays an acquiescent or playful attitude.

Although studies of infant behavior development as presented by McGraw are confined largely to functioning of the motor mechanism, it has also been pointed out that during the first two years of life there are four outstanding periods which may be classified roughly in terms of the type of development taking place most rapidly at the time.

> The first period covers approximately four months and is marked by a diminution of the atavistic reflexes and rhythmical movements characteristic of the newborn. The second period ranges from about four to eight or nine months and is characterized by the development of deliberate or voluntary movements in the superior spinal region and by comparatively reduced activity in the region of the pelvic girdle and lower extremities. The third period is characterized by increasing control of activities in the inferior spinal region and represents the age range from eight to fourteen months. The fourth period covers the remaining ten months and is marked by rapid development in associational processes,

simple or direct, conditional, and symbolic associations, including language (McGraw 1939a).

These broad categories are of practical value, but analyses of behavior development must be explored intensively with respect to both the structural and functional components before the underlying principles involved can be ascertained or formulated. Both theoretically and empirically there is every reason to believe that behavior development moves according to law and order. The determination of these laws of growth awaits improvement in the concepts and techniques of investigation in subject matter of this type. McGraw's studies demonstrate an attempt to order observational data according to a symbolic system, and Weinbach (1937; 1940) and McGraw and Weinbach (1936) have shown how data, even observational data of different physiological phenomena, may be fitted to growth equations. These equations yield constants which are descriptive of the phenomenon of change which characterizes all growth; by means of such equations and constants, features of growth which are qualitatively different can be compared. In view of the propositions set forth by Huxley (1931), Goldschmidt (1938), and Sinnott (1939) concerning the relative rates of growth, it is conceivable that the determination of constants representing the rates of different types of behavior development and the relation between such constants may open the way to a new approach in investigating the process of development in behavior.

DISCUSSION

Certainly among psychologists the term maturation has acquired a somewhat specific meaning, referring essentially to changes in behavior as a result of anatomical or physiological development in the nervous system, and in distinction to changes brought about by exercise or use of the function. It seems fairly evident that certain structural changes take place prior to the onset of overt function; it seems equally evident that cessation of neurostructural development does not coincide with the onset of function. There is every reason to believe that when conditions are favorable function makes some contribution to further advancement in structural development of the nervous system. An influential factor in determining the structural development of one component may be the functioning of structures which are interrelated. Whereas structural growth in such instances is not the product of function directly, neither is it free from or opposed to function as such. Obviously, rigid demarcation between structure and function as two distinct processes of development is not possible. The two are interrelated, and at one time one aspect may have greater weight than the other.

Limited as our knowledge of these two aspects of development is, an understanding of the process of development must embrace studies not only of neural structures and their related functions but also other aspects of the growing organism. In some instances alterations in behavior pattern may be definitely influenced by changes in bone and muscle structures, by hormonal or vascular maturation. For example, not only neural innervation but also the size of the child's hand, the size of the object to be seized, and its mobility will affect his mode of prehension. Changes in static equilibrium, as a result of relative growth in leg length to body height, are factors which affect a child's manner of performance in activities such as climbing and roller skating (McGraw and Weinbach 1939a; McGraw 1939c; Weinbach 1940). There are periods in individual development when emotional adaptability is more easily accredited to a process of physiological maturation than to increased familiarity through repetitive experience (McGraw 1939b).

Maturation is not a process peculiar to certain activities nor is it restricted to early stages of life. Both from embryological studies and from studies of infant behavior developmental evidence has accumulated which suggests that critical periods occur in the growth of any phenomenon when it is most susceptible to definite kinds of stimulation. The critical period for one activity may occur at one time in the life of the individual and at another time for a different activity. The major factor contributing to an alteration in behavior may at one time be the status of neurostructural components, at another time variations in anatomical dimensions, and at another time personal or individual experience. In fact, it probably is the interrelationship of a multitude of factors which determines the course of behavior development at any one time. Determination of the direction of growth in specific behavioral activities, the rate of improvement, and the fluctuations, not so much among individuals, but in the phenomenon per se, constitutes a primary step in investigations of behavior development.

Observations of development in these various disciplines provoke the thought that the "maturation-versus-learning" dichotomy is a cumbersome conceptual framework. In the present state of knowledge a more profitable approach lies in the systematic determination of the changing interrelationships between the various aspects of a growing phenomenon. It has been suggested that relative rates of growth may afford a common symbolic means by which the underlying principles of development may be formulated. Once the laws of development have been determined the maturation concept may fade into insignificance.

❧ 15 ❧

An Open Letter to Parents
of Young Infants

When one ponders the potentials of those ten billion brain cells housed in the small cranium of the infant one is mystified that scientific explorers delayed until the twentieth century to direct their inquiries to the behavior and development of the human infant. Perhaps it was because babies were so common and their behavior was accepted as natural, in no way phenomenal. A study of the universe, the movement of the stars and planets was more challenging.

In a sense this was fortunate for parents of centuries ago. They could bask in their own self righteousness. At times in history parents were granted the powers of life and death over their children, both legal and moral powers. Infanticide was common and in some religions infanticide was sanctioned in sacrificial rituals. While legal infanticide faded, the rights of parents over the discipline and upbringing of their children remained unchallenged. Any formalized system of education for the infant and preschool child was non-existent. The child of these years was expected to grow. The management of young children was in the custody of women, mothers, and nurses. Women were supposed to be endowed with a "maternal instinct" which equipped them with the know-how for dealing with the immature child. There was a philosophical umbrella to protect the women in charge, if the child did not grow according to expectations. If the infant's behavior was deviant or puzzling it was assumed that "he would outgrow it." If he failed to outgrow it "it was God's will." In either case the burden of responsibility did not reside with the mother and she need not develop guilt or any sense of failure on her part.

This state of affairs, a trust in the divine wisdom of motherhood, began to change at the advent of the 20th century. The new sciences, especially psychology, pediatrics, and psychiatry, began to direct their investigative techniques to a study of the "nature and needs" of the child. Books of directives

Unpublished essay, 1969. Included here by permission of Mitzi Wertheim.

257

and guidance for parents began to flow from the popular presses. When faith
in the innate intuitiveness of women was dissipated, a mother's self-confidence was shattered. To try to bring young children up "according to the
book" was vogue. Courses in parent education began to occupy a common
spot in college and university curricula.

SOME PERSONAL REMINISCENCES
AND REFLECTIONS

Back in the early 1920s when I was a graduate student at one of the great
universities an eminent professor spent considerable time arguing the thesis
that the modern woman was going through an evolutionary change which
was destroying her "maternal instincts." A major point of his argument was
that for centuries on end women had relied upon their instincts for rearing
young children, but here in the 20th century they had to take courses and
learn how to do it, just as they had to learn to study for other skills. Now, of
course, the modern woman has just as much potential for developing those
qualities previously referred to as "maternal instinct" as did her great-great-
great grandmothers. But the modern mothers are not equipped with that
great reserve of intuitive sensitivity which served their forebears so effectively. There is no justification in ascribing this difference in intuition to
some alteration of the DNA code.

Factors contributing to this generation difference in mothers are many,
but a few will suffice to illustrate the transformation in the preparation of
girls for motherhood. In the first place, even just a few generations ago, a girl
grew up in a large family—the kinship family—there were always young
growing children around. Older children were involved in the task of caring
for the younger ones. Just by living with them the girl learned to "read" the
signals of the infant and toddler even before the child had attained command
of his native language. In our society, regardless of the size of the family,
children spend a fair share of their waking hours in classes of their peers. Because of the structure of our educational system this compartmentalization
of children on a chronological basis continues until they reach young adulthood. Thus the generation gap begins at kindergarten. There is very little opportunity for the young girl to get a feeling about the way infants and young
children grow. Furthermore during the years of formal education the student
has become linguistically dependent and has blunted that marvelous potential of the young child to respond to a whole multiplicity of stimuli, to make
an interpretation of a living situation and not know how or why it was possible to make that particular judgment. Perhaps that is what is really meant by
intuition. Certainly it is a capacity that young children have. How often does
a mother turn to her three or four year old child and ask what is it the infant

wants or is screaming for. The three year old will be able to tell her though she couldn't pick up the signals herself. One cannot escape the hypothesis that the young child must be sensitive to certain wave lengths of communication which we, as adults, have turned off. Certainly this is an area which has not attracted the exploration of the behavioral scientists. Nor have the educators sought to find a way to cultivate that potential of the young child; they have been too concerned with the development of techniques for cramming facts and logic into the heads of children.

Some of the blame for our present theoretical dilemma goes way back to the Greek philosophers, especially Plato and Aristotle. They began to compartmentalize and categorize behavior, to separate the rational from the non-rational, thinking from feeling. All through the ages of western culture these two facets of behavior have struggled for status and priority. In the American Puritanical tradition, thinking has generally been considered most laudable. The present generation of young people attach great emphasis upon sensory experience. But in daily living they can't be separated. In practically every act they are a mixture. It is the *mixture* and the timing that counts. Without knowing how, the young child seems to be able to convert thinking and feeling into an emulsion. That is the capacity which we need to cultivate and expand; that is a capacity which has been neglected both by investigators and educators. That is the capacity for which we need some new guidelines and new principles if we would better qualify adults for the management and upbringing of young children.

We can credit Plato with another handicap. It was he who took the attitude that until children were able to "reason"—about the age of seven—there was no purpose to get involved in their education. Thereupon six or seven became the customary age for formal education. And that wonderful sensitivity of the infant and young child has been neglected. It was this mystical aura created about the propitious age for thinking and reasoning which precluded exploration into the differences between the learning processes during the early years and the school age. No doubt this type of thinking helped to delay our pursuit of knowledge of the nature of growth and development in the infant and young child.

In the early decades of the 20th century two psychological theories did draw attention to the importance of the early years. For those of us who have had an opportunity to watch the way in which these theories have been adopted or applied in the society it seems clear that they have threatened parental security and self-confidence. One theory associated with the environmentalist contended that by controlling the environment one could make of him whatever one chose him to be. The emphasis was upon intellectual attainments. The burden of responsibility fell squarely upon the parent. The other theory, associated with Freudianism, claimed that adult neuroses were rooted in early childhood experiences, especially parent-child relations, and the emotional involvement therein. Both these theories served to blame the parent for the child's deviant behavior. Parents lost faith in their own intuitive sensitivity. Among the psychologists at large, intuition, like clairvoyance, was an unfit subject for investigation.

Since the scientist ignored this phenomenon of behavior no well defined concepts or theories have been developed comparable to those which have accumulated around the subject of "learning." I like to think of intuition as an "emulsion of sensitivity, knowledge and feeling." It has not disappeared from daily living, despite the disrespect attached to it. How often does one hear himself say, "I just have a feeling thus and so?" How often does a teacher hear a child say, "I know but I can't explain it." The principles or so-called "laws of learning" based upon studies of animals and older children do not explain the phenomenon of intuition.

During this century more factual knowledge about the child has been accumulated than in all the other centuries put together. At the same time we have produced probably the most conscientious yet the most insecure generation of parents at any time in history. They are anxious and insecure because they have lost faith in their own wisdom. Is there any way to restore self confidence to the modern parent? I think there is. But we shall need some new guidelines, perhaps even new cliches. The Baby is our teacher.

SOME SUGGESTIVE GENERALIZATIONS
TO BE LEARNED FROM THE BABY

If babies weren't so common we would all stand in awe, completely bewildered by the fantastic transformations that take place during early development. Just think, in three short years the baby must acquire all of the essential attributes of being a human. He has to gain control of erect locomotion, he has to master enough of his native language to ask the most amazing questions, and with his limited vocabulary he can give his mother a fairly good logical argument within the limits of his own information. It is absurd for us to think that the general principles of learning derived from studies of school children and laboratory animals can adequately explain such incredible transformations in so short a time. We are definitely in need of more adequate theories to cover the kind of learning and development that takes place during the first 3 years.

The most predominant aspect of childhood is growth. Growth means change, a constant flow of change. Healthy child management during these years of rapid growth calls for something more than knowledge.

Even the newborn is a complex bundle of different systems, each interacting with the other. They do not all grow at the same time and at the same rate. Advancing development in one system may alter, even curtail, manifestation in another. For example, advancement in speech may temporarily disturb his early accomplishments in toilet training.

The infant is a complex unit in a complicated, ever expanding system of communications. He both sends and receives signals. The great issue for the parent is to learn to read the baby's signals. This can be a fascinating challenge. Just as soon as the parent has begun to feel confident in detecting a signal the baby may switch to another wave length and change his signal. These changes in signals indicate a new development. Learning to read the signals and the changes in signals and to give the baby a return message "feed-back" so he can know that his signal has been received is a profitable method of cultivating "intuition or sensitivity" in the mother, and in the process will restore some degree of self-confidence to parents.

Consistency in managing the infant and young child doesn't mean sticking to the same procedure or requirements day after day. Consistency within the growth frame of reference means altering one's methods so that they are consistent with the developmental changes which have taken place in the child.

The direction of growth during the early years is in the direction of the child's gaining ever more control over and conscious awareness of his own behavior. It is not the better part of maternal wisdom for a mother to think of herself as the controlling figure. The center of control may shift from time to time in any situation. At one moment the mother may be in command of a situation, but then, all of a sudden, the toddler goes as limp as a wet dish cloth. Where is the center of control then. Actually this capacity to go limp signifies the child's expanding development in gaining control over his environment. When a mother can see these frustrating episodes as signals of the child's development, she is less inclined to be irritated by them.

You may also observe that with the emergence of every new function there is a strong urge on the part of the baby to exercise that activity over and over again. This is your cue to provide an opportunity for that exercise , in other words to ride with the wave of the growing process.

You may also observe the disappearance or diminution of some activity which the infant had been accomplishing quite well. This is the time for you to scan and see if there are partial signs of the emergence of some new activity, perhaps in what seems to be a totally unrelated area.

As one learns to see the fluctuations and rhythms in the process of growth one sharpens one's own observational acuity and cultivates a potential for intuitive sensitivity coupled with knowledge.

When a mother is armed with these skills of observation and interpretation then healthy child-parent relations are off to a good start.

❧ 16 ❧

Challenges for Students of Infancy

Two waves of concern about infancy and early childhood have hit the United States during this century. The first wave occurred during the 1920s and 1930s. We are in the midst of the second. A better appreciation of the current interest can be achieved if we take a glance at some of the parallelisms and divergences which characterized the earlier investigations.

For more than two thousand years, the infant had received scant attention by those engaged in intellectual inquiry (Aries 1965; Sunley 1968). What were the forces that triggered the arousal of researchers to the study of infancy in the 20th century. Certainly among the most influential forces was Darwin's Theory of Evolution. Darwin himself was among the first to conduct systematic observations of his own babies (Darwin 1877). He was not concerned, however, with the nature of infancy or with the process of growth. Darwin's purpose was to accumulate evidence supporting his theory on the origin of the species. His identification of man with all other species provided the theoretical framework for that persuasive dichotomy, heredity *versus* environment. Darwin's theory paved the way for two significant assumptions: (1) that principles of behavior derived from investigations of animal subjects are applicable to the human; and (2) those traits present in the behavior of the newborn child can be ascribed to heredity, and those acquired subsequently are attributable to environmental influences.

The predominant investigators at the beginning of the century were drawn from that rising young academic discipline, Psychology. Those choosing animals as subjects focused their attention primarily on the phenomena of learning. They were of the opinion apparently, that the "laws of learning" were universal, and that neither the age of their subjects nor the species was of prime importance. In fact, many of the animal researchers of that era never bothered to record the age of the rats whose trials and errors were be-

Reprinted from Foreword, *Advances in Infancy Research*, eds. Lewis P. Lipsitt and Carolyn K. Rovee-Collier, Vol 1, Copyright © 1983, Ablex Publishing Corporation, New Jersey. Reprinted by permission of Ablex Publishing Corporation. Reprinted by permission of Mitzi Wertheim.

ing meticulously counted. They did not consider growth and development as variables. Nevertheless, the principles of learning based upon laboratory studies of animals permeated pedagogical thinking and methods during the early decades of this century.

Although not the earliest to make objective observations of infants, John B. Watson was the first psychologist to subject newborn infants to what might be called scientific investigation. His purpose was not to study the process of growth and development. Rather, Watson sought to determine what specific traits, manifest in the newborn, were hereditary and presumably immutable. He found few. He thereupon became the archexponent for the environmentalist. From Pavlov's conditioning experiments with dogs, Watson captured a theoretical frame of reference appropriate for an explanation of complex human behavior. The theory was that if one could adequately control the environment of a growing child, one could, through the process of conditioning, mold that child according to one's choice. Clearly this theory put great responsibility upon adults responsible for child rearing.

After Watson left the academic laboratory and turned to advertising, he published for popular consumption (Watson 1928). Perhaps Watson was the first scientific psychologist to invade popular journalism and to have a direct and immediate impact upon cultural practices in the bringing up of young children. Some of us still remember conscientious parents of the 1930s who would not dare pick up the crying infant; the child might become "conditioned" and cry in order to be picked up. Toilet training was to be started at about three weeks, and children should not be lifted off the potty until they performed, lest the conditioning process be interrupted. Feeding and sleeping were scheduled according to the clock, and parents were admonished not to cuddle or fondle infants lest they become too dependent upon parental attention.

In the laboratories the nature-nurture issue became a prime target for both animal and child investigators. The heredity *versus* environment controversy had them really locked into a conceptual bind. How to determine and to measure that which was innate and immutable, and distinguish this from what was acquired and was amenable to training? Aye, there was the rub!

Gesell challenged the rigid environmental view of Watson (Gesell and Thompson 1929). Gesell contended that there was nothing one could do through training young infants to accelerate their development; one simply had to wait until the cells of the nervous system "ripened." Within this framework, Gesell introduced another dichotomy, that of maturation *versus* learning. Actually, this is closely related to the heredity-environment dichotomy, except that it does imply a kind of unfolding or delayed emergence of innate traits. In any event, it took some of the pressure off parents to start toilet training infants at three weeks, or to assume all the blame for their child's delayed or deviant behavior. On the other hand, another of Gesell's contributions to child development for which he was most famous, would appear to be in contradiction to his maturation theory. It had a pernicious anxiety-arousal effect upon the parental public.

The use of intelligence tests with studies of school children had become accepted by psychologists as an effective means of measuring intellectual potential. The best known of these was Terman's Revision of the Binet which yielded an I.Q. (Intelligence Quotient). The I.Q. was the relation between mental and chronological age as reflected by the test scores. For some time it was taken for granted that the I.Q. reflected innate endowment. Gesell, Thompson, and Amatruda (1934) undertook to set up standardized tests for infants and for children below school age. What can one include in a behavioral test for infants? Only what the child is able to do, such as when he begins to sit, creep, walk, handle objects, and so on. On the basis of ratings on these tests, the child was given a Developmental Quotient (D.Q.), and its correlation with the I.Q. was heralded. While it was never stated categorically by Gesell or his coworkers, the general impression among the parental population was that the infant's D.Q. presaged the potential intellectual endowment. It became the vogue, especially among educated parents, to compare an infant's development against the Gesell norms. It is easy to see how these two contributions of Gesell would have anxiety-arousal qualities among parents. On the one hand, his maturation theory stated there was nothing one could do to accelerate an infant's development. On the other, the child's psychomotor achievements were presumed to forecast mental endowment.

This is not the proper occasion to take up a general discussion of the whole standardized testing movement and its effect upon American society in every facet. It should be pointed out, however, that every scientist longs for a good instrument of measure. The standardized test was the tool which the early developmental psychologists seized upon as a convenient instrument for measuring development. The essence of the standardized tests is that they are notations of achievement. They are like inventories of performances, milestones along the road. But they don't tell *how* the child got there! This focus upon achievements in the 1930s seems to have shunted the experimental psychologists away from the study of the *processes of growth*. It is rather interesting that in the first decades of the century, under the influence of G. Stanley Hall, students of child behavior were called "Genetic Psychologists." Then, as the measurement techniques gained precedence, they came to be known as "Child Psychologists." Within the profession child psychology was always a kind of step-child of general and experimental psychology. Later, for some subtle reason, students of child development came to be called "Developmental Psychologists." But the Developmental Psychologists did not alter their methodologies or concepts appreciably from those that had been practiced earlier by the Child Psychologists. The child was compartmentalized, as the tables of contents of textbooks reveal. These categories went like this: infancy, pre-school, middle years, puberty, and adolescence, or: motor development, language development, emotional development, personality development, and so on. None of these categories is adequate to reflect the process of growth, the phenomenon of constant change, and the interaction of one system with another. Perhaps one reason these aspects of growth and development were not given due recognition by the early experimentalist was that there was no easy method for measuring a constantly changing phenomenon. We should have a specialty called "Growth Scientist."

Research psychologists were not the only ones who called attention to the significance of infancy and early childhood. Nor were they the only ones who contributed to the cultural thinking and practices with regard to child rearing. Perhaps the greatest, certainly the most pervasive, impact upon the culture came from psychoanalytic theories. Freud's ascription of adult neurosis to early childhood experiences lifted somewhat the behavioristic pressure for achievement and placed the emphasis upon personality adjustment, and particularly upon child-parent relationships. It cannot be said, however, that psychoanalytical theories and practices restored parental confidence. It is true that the Victorian taboos over infantile psychosexual behavior such as thumb sucking and masturbatory activities were removed and parents were admonished to give love to their child. Parental love became something of a panacea. Parents were warned of the dangers of rejection or over-protection of the child. In practice these warnings were often confusing. Was a mother overtly demonstrating love in order to cover up her basic feelings of child rejection? Certainly during this century a great body of factual data about child development has accumulated and a plethora of concepts and theories have become a part of the cultural milieu. Yet we still have a generation of uncertain, anxiety-ridden parents.

THE CURRENT FOCUS ON EARLY CHILDHOOD

It is not possible to pinpoint any particular ideologies or theories which have given rise to the present interest in early childhood development. The forces were many; they were complex and intertwined. Sputnik shocked the nation out of a state of educational complacency. The disparity of educational opportunities and achievements of children from differing socio-economic and ethnic groups was brought to light. It was suggested that children from less favorable environments entered school with their educational handicaps already established. To alleviate this situation, the Federal Government set up Head Start programs. The outcome of the Head Start programs has led to the claim that even the prekindergarten period is too late. Education begins in the cradle.

Clearly the goal of this current wave of concern is to develop the optimum potential of all children. The prevailing notion is that these goals can be achieved by manipulation of the environment in which the child lives. To some extent these ideas are reinforced by experiments on the effects of "sensory deprivation," "prolonged isolation," and the comparative effects of "enriched and impoverished" environments. Such studies have been conducted on animals, children, and adults. Once again, the emphasis seems to be shifting to the environmental side of the continuum, but it is not locked in with the old heredity-environment dichotomy. It is generally recognized now that nature-nurture are interdependent forces and to try to separate them clouds inquiry. A few studies (McGraw 1935; Moore 1960; Fowler 1962) have demonstrated that the performances of the young, in some specific activities, can be advanced beyond normal expectancy. But we have not as yet learned how to develop to the maximum the potential of all growing children.

CHALLENGE FOR THE
RESEARCHERS OF GROWTH

Contemporary Growth Scientists are the legatees of a vast body of concepts, theories, and research strategies inherited from the "psychological establishment." Of course, the Growth Scientists will be drawn from many disciplines and from diverse areas of psychology including Developmental Psychology. Already it is apparent that some dyed-in-the-wool experimentalists are selecting the human infant in preference to animals for special investigations. The challenge for all the students of growth, regardless of their scientific expertise and theoretical orientation, is to scan the legacy of knowledge and skills, and to have the courage to rule out those theories and techniques which are not applicable to the study of complex, ever-changing phenomena relating to the growth and development of organisms. Many experimentalists fail to consider that their own preconceptions may operate as uncontrolled variables within a particular situation.

Will the experimentalist, skillful in the manipulation of the variables and instruments of measurement, be able to recognize that the way the infant is held or positioned may also be a factor in the results obtained? Will the examiner be so focused on toddlers' responses to the items set before them that it goes undetected that the child's wiggling and climbing off the chair and running toward the door is his way of saying that there is pressure on his bladder? Will researchers trained to use the I.Q. be able to devise strategies to evaluate a multiplicity of systems constantly in flux, each system influencing another and in different degrees? All growth and development is not in the form of accretion. The Growth Scientists will need to design methods which reveal the rises and falls, the pulsations and rhythms manifest in the growth of a given function. An understanding of these pulsations and rhythms may become promising guidelines for the development of optimum potentials of the growing child. There is evidence that many current investigators (see Endler, Boulter, and Osser 1968) are alert to the problem and that is the first step to improving methodologies.

THE CHALLENGE OF CULTURAL ACCEPTANCE OF SCIENTIFIC THEORIES

In the past it has been traditional for scientists, especially those dealing with basic sciences, to be removed from the applied implications of their findings. They were searching for fundamental truths, and whatever society did with them was none of their concern. On the other hand, many atomic physicists have voiced a sense of responsibility for the way society makes use of their knowledge. During this century, we have been able to see how many psychological theories have been applied and misapplied in the matter of child rearing and education. If the periods of infancy and the early years are as important for total development as generally contended, it is reasonable to expect the behavioral scientists to take some responsibility for the way in which their thoughts and theories are adopted into the cultural patterns of child management. Just how this can be done is not clear because it has never been systematically undertaken by any scientific discipline. The general public has faith in science, and the mass media is quick to announce that science has proved one thing or another. Sometimes the misapplication of a theory may be ascribed to the use of a particular word, perhaps a word that was originally drawn from another discipline.

Let us consider for a moment some current thoughts which have the potential for creating parental anxiety. Take the question of "critical periods" as applied to learning. The concept was first used by embryologists. It was reinforced by Lorenz's (1935) study of imprinting. Recently it has been emphasized in connection with the studies of effects of an impoverished environment. It has been asserted that if the impoverishment occurs at critical periods, then the damage done may be irreversible (Bowlby 1969). Back in 1935 (McGraw 1935), the writer applied the term "critical period" to the acquisition of motor skills during infancy. If the agreed meaning of the term "critical periods" carries the idea that whatever is attained in development or learning *must* be achieved during a specified period, then the term should not be applied to normal behavioral growth. In the aforementioned instance, it was intended to signify that there are *opportune* times when specific activities can most economically be learned. If that opportune time is missed, then the methods of instruction should be altered for later learning of the same function. It is the irreversibility of damage done which adds emotion and fear to the "critical period" concept.

The amount of emphasis attached to certain concepts can also distort their meaning when adopted into the culture. Take, for example, the current emphasis on cognition. No investigator would contend that cognition operates independently of other aspects of learning. Yet merely because it is the focus of investigative activity, cognition, like the old notion of personality adjustment, is a kind of umbrella for other goals. Expose the child to the right knowledge, in the right way, and at the right time, and the job would be well done.

Perhaps most urgently of all, Growth Scientists need to review the accepted principles of learning as they have been articulated and generally accepted. These learning principles were determined largely by animal studies in laboratory situations and studies of children in the classroom. There is every reason to suspect that they are not applicable to the process of growth taking place during infancy and the early years. There is a pressing need for totally new guidelines for the benefit of those persons responsible for the management and socialization of the child from birth to three years. The most dominant force is change, change in the organism and change in behavior from day to day. Consistency in parental management doesn't mean setting up a pattern or rule and sticking to it. It means dealing with a child in a manner consistent with the individual's developmental changes. To do this effectively requires knowledge, sensitivity, intuition, and flexibility. So the challenge is to orient mothers and teachers toward the concept of change, not toward stability in the ordinary sense. Parents should be taught to observe, to scan, and to detect the non-verbal as well as verbal signals of child growth, and to design methods of instruction accordingly.

The United States may well be at the threshold of institutional reorganization for the care and education of the young. To help children develop will require new knowledge and special preparation on the part of those responsible for them. They need to be knowledgeable, but also intuitive and observant. Child care specialists and parents will require preparation quite unlike that offered to elementary school teachers or even mothers of today.

The Growth Scientists are challenged to provide a theoretical frame of reference for the education of infants and young children. They are advised also, to take account of the way in which their theories and pronouncements are adopted into the culture so that the growing child of today can confidently meet the social changes of the twenty-first century.

Epilogue: A Developmental Neurologist's "Homage"

BERT C.L. TOUWEN

My contact with Myrtle McGraw took place in the beginning of the sixties, when I got hold of her book on neuromuscular maturation. I read her studies on Johnny and Jimmy, and I was excited. I realized two things, which can be brought into one formula: Observation of normal variation. That is what McGraw taught me.

Myrtle McGraw was one of the first who emphasized the value of the "natural experiment": observation of normal development in normal infants and children, instead of experiments with specifically lesioned animals. Moreover, she was one of the first who paid attention to variation of development as characteristic for normality.

The importance of the first point—observation of intact normal young infants and children as the proper means to find behavioral correlates to morphological brain maturation—becomes clear if one realizes that McGraw got her education in a time that was dominated by the reflex paradigm developed by the famous neurophysiologists of the beginning of this century. Sherrington (1906) had formulated the concept of the reflex as a model to explain brain activity, although, to be honest, in his famous book he queried the value of the reflex as the only or major explanation of the nervous system's activity: in his opinion the reactions of a healthy organism were too complex for such a simple explanation. This was the fascinating period in which in laboratories much knowledge was gained about brain function, based on his concept. Specific ablations in (adult) animals led to insight in various brain mechanisms, especially with regard to what were considered to be basic reflexes and reactions for postural activity. In 1924 Rudolf Magnus described placing and righting reactions such as could be found in experimentally lesioned adult animals, in young, intact animals, and Georg Schaltenbrandt did the same in human infants in 1925. Schaltenbrandt considered the disappearance of these reflexes to be based on inhibition of (mainly) red nucleus activity by the maturing corpus striatum (basal ganglia). Rademaker (1931) considered the reflexes as a kind of basic units, which were integrated into complex coordinated motility during development. In his view the reactions as such remained present, and could reappear under specific circumstances, e.g., disease or old age.

The apparent presence of such primitive reflexes and reactions labelled the newborn baby and the young infant as primitive or sick adults. Other examples of reflexes thought to be present in all normal infants though only in sick adults, are the asymmetric tonic neck reflexes, grasp reflexes, and the Babinski reflex of the big toe—although in the case of the latter, Babinski (1898) himself considered the dorsiflexion of the big toe in infants as different from that in the adult patient, arguing that it would not do to regard a healthy baby as a disordered grown-up. A. J. Capute et al. still stated in their preface to a booklet published in 1978 with the title *Primitive Reflex Profile* that term infants

> are born with a tendency to a decorticate posture. Prematures present with an extensor posture resembling decerebrate rigidity. This evolution from 'physiologic decerebrate', as seen in the premature, to 'physiologic decorticate' as seen in the term newborn, is an interesting phenomenon. This clinical observation, along with the extremely variable tone recorded in the first half year of life, has led to the concept of 'physiologic newborn extrapyramidal cerebral palsy' as a transitory stage of normal motor development.

Such a conception considers the infant as a primitive adult and development as a kind of recovery of disease. Still, it can be seen as an example of consistent reasoning starting from the neurophysiological conceptions of the first half of this century. Another example is Andre Thomas' point of view in the fifties. Andre Thomas turned to the study of the newborn after a long life in adult neurology. Influenced by his "adult" conceptions he described the increase of muscle power of the 2 month old infant as "physiological hypertonia," and used terminology such as "physiological astasia and abasia" for the description of the first display of the hesitating way of standing in healthy infants (Andre Thomas and Saint-Anne Dargassies 1952). Obviously, such concepts should be obsolete, nowadays. The infant's brain is not a primitive edition of that of the adult. The infant avails of a nervous system, that is both morphologically and functionally age adequate. His nervous system might be primitive, when it is diseased or damaged. The so-called "primitive" reflexes might indeed be primitive in a baby with a damaged brain. In that case asymmetric tonic neck reflexes (as an obligate or imperative response on turning the face to one side; obligate meaning that the infant cannot overcome the postural reflex without moving his head) stereotyped grasp reflexes, which resemble adult "forced grasping", or a stereotyped tonic dorsiflexion of the big toe with fanning of the other toes may occur. Normal infants show "early infantile" reactions, which are too variable and too complex to be called primitive reflexes (Touwen 1984).

Although she did not state it in so many words, McGraw smelled as it were, this paradigmatic change in interpreting infant reflexes. She remained largely within the realm of the reflexological paradigm which prevailed in the period of her trainig as a scientist, but she realized that normal development can only be studied in normal individuals and that normal infants are not sick adults. She recognized that normal development cannot be a product of conditioning through, and inhibition of reflexes. Moreover, she found that extrapolation from animals to humans is risky, and so she discarded the neurophysiological and neurological knowledge about both adult patients and specifically lesioned animals as a source for an infant's development. She turned to the baby itself.

McGraw was not a neurologist. She was a developmental psychologist when that discipline was in its infancy. She was not interested in reflexes and reactions from a neurological point of view. She focused on normal motor functions, realizing that all kinds of reflexes and reactions, if essential for proper neural functioning will be represented by the sensori-motor activities. She wanted to relate normal development of function with what was known of the structural maturation of the brain. She looked for structure-function relationships in normal organisms, reflecting that normal and abnormal organisms are widely different, and that good knowledge of normal function is a prerequisite for the understanding of abnormal function. And she was struck by the presence of variations. Perhaps the most characteristic feature of Myrtle McGraw's scientific attitude—the feature that makes her a truly great scientist—is her ability to keep her eyes and mind open for new or unexpected things, and the ability to accept them and to try to understand them. Of course she was not the first one to see that children differ and that the developmental course of infants may vary. But she saw, and appreciated, the occurence of intra-individual variations. She realized that development is not a continuous process, but that relapses in function occur, or accelerations and decelerations. And, above all, she perceived the child as a being in its own right, and not as a small edition of an adult.

The appreciation of variations in the developmental course of individual infants and of individual motor functions in infants (and their developmental sequences) as a characteristic of normal development has been the second great contribution of McGraw, at least to my education in the developmental neural sciences. She pointed out the wide overlap between the various consecutive phases which can be distinguished during the development of single motor functions. McGraw believed strongly in growth, that is, in the overall progressive trend of development, but she also believed in variations between and within infants. I have shown how the overlapping developmental phases contribute to an apparent inconsistency and variability of development: If phases overlap, it may happen that the observer sees only one during his assessment, and this is not necessarily the most advanced one (Touwen 1971). It stands to reason that this state of affairs affects the strength of interrelationships between developmental courses of different motor phenomena during infancy (Touwen 1971). Here again McGraw showed her courage. Arnold Gesell considered normal development to be a rigid and predictable process. Some variations could occur, based on differences in genetic make-up, but these should be small and mainly inter-individual. A deviation of the established developmental course, and in particular if the variations were intra-individual, suggested an abnormality (Gesell and Amatruda 1947). Also Illingworth (1966) warned against unexpected variations in developmental sequence, because usually there is a specific cause for them. Bayley (1969) merely mentions differing rates of development, especially during the first months of life, but she doesn't seem to appreciate their significance. Andre-Thomas admitted that there may be inter-individual differences in developmental rate and sequence, but these would be exceptions to the rule (Andre-Thomas and Saint-Anne Dargassies 1952). Textbooks on child neurology are still a bit reticent about variations during development. This is understandable, for so far, extreme deviations in rate and sequence, especially when occurring in many sensori-motor functions—spontaneous and voluntary activity or reactions and responses—at the same time or in the same way, can be indicative of abnormal development (Nelligan and Prudham 1969; Touwen 1978).

McGraw was very prudent in her considerations. Her data on the development of normal infants in her book, *The Neuromuscular Maturation of the Human Infant,* clearly show both inter- and intra-individual variations. She commented on them carefully. Although her standpoint is evident with statements such as: "the variability of each child throughout several behavioral activities". ... "The centers which relate to the reflexes involved in early sitting, creeping, and walking behavior are different and relatively independent. ... "The small amount of association within individual infants" (McGraw 1943:117–118). She remained true to the concept of consistency of development which was prevalent in her days. But she realized that the organization of the brain may differ from area to area. As she said in 1985: "My concept was of multi-systems developmental processes emerging and advancing at different times and different rates, but finally interacting, integrating, and synthesizing for the creation of new performances or traits" (McGraw 1985:170). Indeed she remarked on the "losses and regressions in performances" in her "Lecture to the Laity" in 1943 [see Chapter 11]. But it was left to later researchers to show what inconsistency of normal development looks like, progression and regression going hand in hand, both morphologically (e.g., neuronal cell death, dendrite and axone retraction, synapse elimination as normal maturational processes) and functionally (e.g., inconsistencies or relapses in function during periods of transition). The latter functional events were anticipated by McGraw when she noticed the "disorganization" occurring between consecutive developmental phases.

It is not my intention to discuss the obviously time-related ideas which lie at the basis of many of McGraw's interpretations. Her association with Frederick Tilney, later LeRoy Conel, and her being influenced by Coghill's concept of individuation are well known. It was inevitable that McGraw used the traditional dichotomy between cortical and subcortical generation of behavior and that she was heavily involved in the nature or nurture discussion. On the one hand, knowledge about developmental neurobiology was still in its infancy in her time, on the other hand, she could not avail of our sophisticated technological means of examination and observation, such as EEG and ultra sound registration of the motor behavior of the fetus *in utero*. Especially the use of ultra sound has changed thoroughly our ideas about the cortical and subcortical shares in the infant's behavior. Gradually a hierarchical model of brain function, in which the cortex occupies the top position, has been replaced by a more interactional model, in which variable brain areas can take command, depending on the task in hand (e.g., Mountcastle's 1978 "changing command centers"). An attractive neurobiological basis for the ontogeny of the relevant brain mechanisms has been forwarded by Purves (1988). According to his "trophic theory" synaptic activity plays an essential role in the establishment and the operationalizing of neural circuitry. Yet, this theory does not explain the fundamental capacity of the developmental brain to switch on and off particular neural activities at will, such as the capacity of the normal infant to grasp voluntarily, implying the ability to loosen the grip, in the presence of the palmar grasp reflex (Touwen 1987).

From pre-term age onwards signs of cerebral cortical activity can be found, although obviously qualitatively different from that in the older infant. The specific contribution of this cortical activity to fetal motility is not known, but in its absence spontaneous movements suffer. They are less variable, but awkward and slow. We know now that from the seventh or eighth week of pregnancy onwards the fetus shows actively generated motor patterns (DeVries, Visser, and Prechtl 1984; Prechtl 1989). That is the same age at which Davenport Hooker (1952) could elicit reflexes for the first time. But we realize now that Hooker's fetuses, who were obtained from medical abortions, were *ad finem,* and therefore more comparable to deeply comatose patients than to healthy unborn infants. I gather that the capacity of the nervous system to generate active motor behavior besides its capacity of reactivity is well established nowadays. This double capacity can be used for the evaluation of fetal well-being: In case of danger, e.g., when severe intra-uterine growth retardation is present, and/or intra-uterine asphyxia which threatens fetal life, changes in the quality of the fetal spontaneous activity may be the first sign of the fetal stress (Bekedam 1989).

At about twenty weeks gestation the fetus avails of all motor patterns which can be found in newborn infants. We think that the second half of pregnancy is used for a complex process of calibration of motor activity, in which the fetus "learns" how to use his or her muscular system. After birth infants incorporate the necessary activity for counteracting gravity. Indeed we find a complex process of progressive changes in motor patterns and motor quality during the first months after birth, during which the originally dominating rather large, slow, and rather ungraceful writhing generalized movements of the newborn are replaced by the small, gentle, graceful, and short lasting so-called fidgety movements of the ten to twelve week old infant (Hopkins and Prechtl 1984; Hadders-Algra and Prechtl 1992). It seems that the presence of the latter type of movements reflects a developmental level from which voluntary motor activity can make its appearance. In the mean time the visual system has matured, which is reflected by the stabilization of fixation, good conjugation of eye movements and accurate pursuit movements. The coupling of the two systems—motor system and visual system results in the establishment of eye-hand coordination. The spectacular changes of the quality of the movements during the first postnatal movements indicate a major transition of brain function (Prechtl 1984). They are corroborated by specific EMG changes (Hadders-Algra et al. 1992) and that may be important from a therapeutic point of view. Systematic evaluation of the age-specific changes can help with the selection of infants who need follow up as they are considered to be at risk for developmental disorders. If we can clarify how, in case of abnormality, changes arise in the organization of sensori-motor activity, means may be found to influence this deviant development. A combined *transactional* effect of nature and nurture is shown by the finding that so-called low risk pre-terms show a somewhat different developmental course of their generalized movements until fidgety age (Hadders-Algra 1993). Possibly this results from the fact that the premature infant has to incorporate gravity in his proprioceptive calibration of muscle activity at a moment that he is still fully occupied with intra-uterine calibration under quite different gravity conditions. Moreover, the premature has a larger number of degrees of freedom for direction and speed of movement, as the constraints due to the intra-uterine position have disappeared.

Heinz Prechtl (1977) has designed an age-adequate neurological examination for the newborn, based on the concept of the age-adequate brain. This concept states that at every age the normal brain is equipped with the capacities required for proper functioning at that age. Knowledge of age-adequate functional display of the nervous system will enable the examiner to recognize abnormalities in the case of disorders. I have described the neurological development during infancy in a group of carefully selected so-called low risk infants, who were longitudinally followed until the age of walking without support. McGraw's example of proper descriptions of the motor activities, and a full appreciation of the presence of variations between but also within infants had a steady and beneficial effect. I was able to expand in particular on the issue of the variations during normal development (Touwen 1976). Besides the well known inter-individual variations also intra-individual variations in age of appearance (or disappearance) or speed of development of the several items (spontaneous and voluntary activity or reactions and responses) turned out to be present. And within functions the various phases of development that can be distinguished were of variable length, again between and within infants. Finally the majority of items showed regressions of scores—a relapse of function—which we have called inconsistencies. Such inconsistencies occurred in all infants and in most items, but to a varying extent both in infants and in items, without any interrelationship. Of course obvious reasons for regressions in function such as accidents, diseases or sudden events had been excluded, the inconsistencies would have shown more interrelationships in that case anyway. We interpreted the inconsistencies as a reflection of morphological changes in the nervous system during maturation: the production of connections, presumably mainly between intra-cranial structures and spinal cord mechanisms. In our opinion inconsistencies belong to the characteristics of the normal development of the nervous system. They seem to be epiphenomena which occur when the system regresses somewhat in order to be able to generate new modes of motor behavior. The most spectacular example is the pointing phase which occurs between the scissor type and the pincer type of voluntary grasping, and which represents a short lasting but thorough regression in-as-much as the infants apparently are not able to grasp at all during this short phase. Pointing may occur also during the pincer grasp period, more rarely together with the scissor type of grasping, but the exclusive pointing phase usually doesn't last longer than about half an hour or an hour. Mothers report often unusual and fussing behavior during this period as if the infant suffers by its incapacity. Morphologically it appears that during this period of pointing synapses are formed between corticospinal fibers and cervical motoneurones (Touwen 1976).

This work on neurological development in infancy culminated in the concept of variability as a characteristic of normal development: The ability of the nervous system to vary warrants the development of manifold strategies for every function, and this is a pre-requirement for purposeful, efficient and adaptive functioning. The diagnostic significance is obvious: A lack of the ability to vary will inevitably lead to a lack of adaptive (both active and passive) capacities. Examples are the various types of cerebral palsy, in which the lack of sufficient and proper strategies has lead to non-adaptive and stereotyped motor behavior, and some types of clumsiness.

Two types of variability are distinguished. During infancy primary, or indiscriminate, variability prevails: The infant develops manifold strategies for all motor functions. After infancy, let us say from the age of walking free onwards, we speak of toddlers. During toddler-age the development of secondary, or adaptive, variability takes over. Development becomes qualitative rather than quantitative. The number of motor functions doesn't increase very much, but the way in which they are performed change so impressively that this secondary stage can be described as a transition of neural function. Especially between 2 and 3 years the character of the motor behavior changes rapidly in two respects: In the first place the performances proper become more fluent and smooth in appearance; movement patterns become more differentiated, that is to say, the movements seem to consist of many detailed and carefully integrated parts in contrast to the more global and rough appearance of the same motor patterns in younger children. Apparently the child has mastered the ability to unravel, as it were, the large movements in their composing parts, and at the same time build up the movements with help of these parts into an integrated and purposefully adaptive movements. In the second place the child develops the ability to select the preferred strategy of the moment from the large number of strategies collected during infancy, and he automatizes the chosen strategy for the particular performance. At the same time he keeps the capacity to recruit some other strategy for the same purpose, if needed. The motor behavior of many psychomotor retarded patients, some types of clumsiness (i.e., those types in which no specific neurological signs can be found) and hyperactivity are examples of a delayed or deficient development of this second type of variability.

McGraw pointed already to the training aspect of the repetition of newly discovered movement patterns. Observation shows that this repetition is not mere copying from the one time to the next; with every performance the infant introduces tiny alterations (Touwen 1994). In this way he expands his number of strategies for that particular performance and he uses them indiscriminately. During toddler-age the child appears to select which mode of movement fits his present need best. At the same time functions become operational which had developed already, but which were not yet used routinely. An example is rotation of the trunk during sitting. This rotation can be elicited from about six to eight months onwards in the majority of infants, but hardly anyone of them shows it during spontaneous activity. Older infants and young toddlers use rotation to turn around on their buttocks without moving their shoulder-and pelvis-girdle differentially. From about two or two and a half years onwards rotation of shoulders on pelvis forms an automatized part of fluent and efficient trunk movement (Hempel 1993a; 1993b).

Primary or indiscriminate, and secondary or adaptive, variability are both properties of the infant's nervous system. Both are genetically and environmentally determined, that means that one infant may be earlier, faster or "richer" than another infant, and that environmental circumstances may have a negative (disease, damage of the brain) or a positive (stimulation, "rich" environment) effect. As said, primary variability denotes the ability to develop variations which are not yet necessarily applied purposefully. Toddlers are characterized by the development of secondary, or adaptive, variability, denoting refinement of motor patterns, the ability to select the proper variation at the proper time, and to automatize this variation accordingly, and the ability to combine motor patterns if required. The actual background of these properties and their interrelationship is far from clear. Does an extensive primary variability lead to an efficient adaptive variability? Has adaptive variability something to do with "intelligence" and maturation that makes adaptive variability a precondition for learning, and not merely for motor cognition? My co-worker, Rietje Hempel, has shown that the various qualitative aspects which can be described are weakly, or not at all, correlated. The developmental rate during infancy is not correlated with the development of adaptive variability either. For instance, the age of walking free is not related to the development of qualitative aspects of walking. A child who has walked for a longer period because he was an "early walker" does not necessarily show a more mature and more adaptive walking pattern at, say, two years, than a child who just started to walk without support (Hemple 1993a). Apparently we have to accept that neural mechanisms, which have been shown to develop independently rather than in close interaction during infancy (Touwen 1976), keep up this independence during toddler-age. It is attractive to presume that the changes in synapse density which have been described in many cortical areas, and which reach a peak during this same period, form a morphological counterpart of these functional changes (Rakic et al. 1986). But actually the relation of this morphological process with functional development and later outcome is still largely unknown.

McGraw's intention was the unraveling of brain function/brain structure relationships and she used new and rather unorthodox methods for attaining her goal. She formulated her ideas carefully; although she stuck to the paradigm of her time which emphasized the reflex background of the young infant's motor behavior, she endeavored to distance herself from the "pure" reflex conception and to turn to the active brain, in which she wanted to distinguish cortical from subcortical components of the infant's repertoire. She realized that cortical and subcortical functions interact throughout development, but, in line with the traditional conceptions, she saw the increasing cortical influence during infancy as mainly an inhibiting one. We do not favor this terminology any longer, on the one hand because we believe that a distinction between cortical and subcortical is not very helpful, on the other hand because cortical activity involves more than merely inhibition. At any age the infant's central nervous system works as a whole, and is operational in all its parts, be it that its complexity, and therewith its displayed activity is age-adequate, and quite different from that of an adult. That is not to say that there is no change in the cortical component's role in so far that can be separated from the rest of the brain's activity; from the second postnatal month onwards cortical areas become increasingly important for the initiation and planning of motor (and other) performances. Inhibition will naturally play a role, and an essential one. But it remains very difficult to disentangle a specific "cortical" (which parts?) and specific subcortical (which parts?) share in all this.

Still, as far as specific function/structure relationships are concerned, we do not know much more than McGraw did. We know much about neurological development in various animals. But we have realized that the feasibility to extrapolate from findings in animals to developing human babies and children is limited. We know something about fetal motility, but we are not sure if and how it relates specifically to postnatal motor function. The question whether fetal movements should be considered as pertinent precursors of specific postnatal motor activity is still open. Perhaps the question is too teleological: It is possible that fetal motility is mainly important for its own sake (e g., proprioceptive calibration), and for its own age (e g., for the proper development of skin and circulation, for skeletal and muscular development, for lungs and intestine). Of course function is essential for proper neural (and not only neural) development and there the circle closes. Our knowledge about human brain development in relation to functional development is still very small. And yet, I expect that the near future will bring some surprises. Technological developments which could not be dreamt of when McGraw started have opened new perspectives. In that regard I think of noninvasive techniques such as nuclear magnetic imaging and spectometry, or magnetic encephalography, or PET techniques. The time will come that we can apply such new techniques in infants without the present hesitations. Then perhaps we will be able to see what the brain does at the moment that the infant performs.

McGraw's point of view in respect to the nature/nurture issue has been largely confirmed. As Ruse (1993) tells us: "No biologist worth his or her salt would deny that all organic features, most especially behavioral features, are a function of the genes interacting with the environment." She fought the pure maturational (read: nature) background of development which was postulated by such researchers as Gesell. She sought to disprove the "tabula rasa" concept of the Behaviorists' views which were rather dominant at the time. Nowadays we know that function is essential for the proper development of structure, though structure is required for the emergence of function. I suggest that the formulation that nature presents the boundaries between which nurture can play (freely adapted from Waddington 1975) would be acceptable to her.

It seems perhaps a long way from a developmental psychologist who started to work in the thirties to a developmental neurologist working in the nineties.

Actually the way is not so very long. The neurologist grasped what he could use and went on where the psychologist left off. And clearly the similarities in attitude were more important then the apparent difference in profession. A short essay of memories, published after her death, ends with a typical McGraw (1990) remark—a typical scientist's remark—"Anyway, it's worth the try." Well, it was worth the try.

References

Andre-Thomas; S. Saint-Anne Dargassies. 1952. *Etude Neurologiques sur le Nouveau-Ne et le Jeune Nourrissom.* Paris: Masson.

Babinski, J. 1898. "Du phenomene des Orteils et de sa Valeur Semiologique," *La Semaine Medical* 18:321.

Bayley, N. 1969. *Bayley Scales of Infant Development.* New York: The Psychological Corporation.

Bekedam, D.J. 1989. "Fetal Heart Rate and Movement Patterns in Growth Retardation." *Ph.D. Thesis.* Groningen University, the Netherlands.

Capute, A.J., P J. Accardo, E P.G. Vining, J E. Rubenstein, and S. Harryman. 1978. *Primitive Reflex Profile.* Baltimore: University Park Press.

De Vries, J.I.P., G.H.A. Visser, and H F R. Prechtl. 1984. "Fetal Motility in the First Half of Pregnancy." In *Continuity of Neural Functions from Prenatal to Postnatal Life. Clinics in Developmental Medicine, no. 94,* ed. H.F.R. Prechtl. Oxford, England: Blackwell Scientific Publications Ltd.

Gesell, A. and C.S. Amatruda. 1947. *Developmental Diagnosis.* New York: Harper and Row.

Hadders-Algra, M. 1994. "General Movements in Early Infancy: What do They tell Us About the Nervous System?" *Early Human Development forthcoming.*

Hadders-Algra, M. and H.F.R. Prechtl. 1992. "Developmental Course of General Movements in Early Infancy. I. Descriptive Analysis of Change in Form." *Early Human Development.* 28:201–215.

Hadders-Algra, M., L.A. van Eykern, A.W.J. van den Klip-Nieuwendijk and H.F.R. Prechtl. 1992. "Developmental Course of General Movements in Early Infancy. II. EMG Correlates." *Early Human Development* 28:231–253.

Hempel, M.S. 1993a. *The Neurological Examination for Toddler-Age.* Ph. D. Thesis. Groningen University, the Netherlands.

———. 1993b. "Neurological Development During Toddler's Age in Normal Children and Children at Risk for Developmental Disorders." *Early Human Development.* forthcoming.

Hooker, D. 1952. *The Prenatal Origin of Behavior.* Lawrence, Kansas: University of Kansas Press.

Hopkins, B., and H.F.R. Prechtl. 1984. "A Qualitative Approach to the Development of Movements During Early Infancy." In *Continuity of Neural Functions from Prenatal to Postnatal Life. Clinics in Developmental Medicine, no. 94*, ed. H.F.R. Prechtl. Oxford: England: Blackwell Scientific Publications Ltd.

Huttenlocher, P.R., de C. Courten, L.J. Garey, and H. van der Loos. 1982. "Synaptogenesis in Human Visual Cortex: Evidence for Synapse Elimination During Normal Development." *Neuroscience Letters* 33:247–252.

Illingworth, R.S. 1966. *The Development of the Infant and Young Child, Normal and Abnormal.* Edinburgh, England: Livingstone.

Magnus, R. 1924. *Korperstellung.* Berlin: Springer.

McGraw, M.B. 1943. "Let Babies be Our Teachers." *March of Medicine* ed. New York Academy of Medicine. New York: Columbia Press.

———. 1943. *The Neuromuscular Maturation of the Human Infant.* New York: Hafner (repr. ed. 1969).

———. 1985. "Professional and Personal Blunders in Child Development Research." *The Psychological Record* 5:165–170.

Mountcastle, V.B. 1978. "An Organizing Principle for Cerebral Function: The Unit Module and the Distributed System." In *The Mindful Brain.* eds. Edelman, G.M. and V.B. Mountcastle. Cambridge, Mass.: MIT Press.

Neligan, G. and D. Prudham. 1969. "Potential Value of Four Early Developmental Milestones in Screening Children for Increased Risk of Later Retardation." *Developmental Medicine and Child Neurology* 11:423.

Prechtl, H.F.R. 1977. *The Neurological Examination of the Fullterm Newborn Infant.* 2nd. revised and enlarged edition. London: Heinemann

———. 1984. "Continuity and Change in Early Neural Development." In *Continuity of Neural Functions from Prenatal to Postnatal Life. Clinics in Developmental Medicine, no. 63*, ed. H.F.R. Prechtl. Oxford, England: Blackwell Scientific Publications Ltd.

———. 1989. "Fetal Behavior." In *Fetal Neurology. International Reviews of Child Neurology.* eds. A. Hill and J. Volpe. New York: Raven Press.

Purves, D. 1988. *Body and Brain.* Cambridge, Mass.: Harvard University Press.

Rademaker, G.G.J. 1931. *Das Stehen.* Berlin: Springer.

Rakic, P.J.P. Bourgeois, M.F. Eckenhoff, N. Zecevic, and S. Goldman-Rakic, 1986. "Concurrent Overproduction of Synapses in Diverse Regions of the Primate Cerebral Cortex." *Science* 232:232–235.

Ruse, M. 1993. *The Darwinian Paradigm.* London: Routledge.

Schaltenbrandt, G. 1925. "Normale Bewegungs-und Lage-reaktionen bei Kindern." *Deutsche Zschr. fur Nervenheilkunde/* 87:23–59.

Sherrington, C. 1906. *The Integrative Action of the Nervous System.* New Haven: Yale University Press. Reprint. ed. 1961.

Touwen, B.C.L. 1971. "A Study on the Development of Some Motor Phenomena in Infancy." *Developmental Medicine and Child Neurology* 13:435–446.

———. 1976. "Neurological Development in Infancy." *Clinics in Developmental Medicine, no.58.* London: Heinemann.

———. 1978. "Variability and Stereotypy in Normal and Deviant Development." In: *Care of the Handicapped Child. Clinics in Developmental Medicine no. 67*, ed. J. Apley. Oxford, England: Blackwell Scientific Publications Ltd.

_____. 1984. "Primitive Reflexes-Conceptional or Semantic Problem? In *Continuity of Neural Functions from Prenatal to Postnatal Life. Clinics in Developmental Medicine, no.94,* ed. H.F.R. Prechtl. Oxford, England: Blackwell Scientific Publications Ltd.

_____. 1987. "Neurological Development of Infants: Structure-Function Relationships and Their Significance for the Early Detection of Deviations." In *Neonatal Brain and Behavior,* ed. H. Yabuuchi, K. Watanabe, and S. Okada. Tokyo, Japan: University of Nagoya Press.

_____. 1994. "How Normal is Variable, or How Variable is Normal?" *Early Human Development.* forthcoming.

Waddington, C.H. 1975. *The Evolulion of an Evolutionist.* Ithaca, New York: Cornell University Press.

Bibliography of References
in Essays by Myrtle McGraw

Anderson, J.E. 1932. "Emotional Development and Adjustment." 12th Ohio State Conference, Columbus, Ohio (September).

Angell, J.R. 1908. *Psychology: An Introductory Study of the Structure and Function of Human Consciousness.* New York: Holt.

Angulo y Gonzalez, A.W. 1930. "Neurological Interpretation of Fetal Behavior: The Progressive Increase of Muscular Activity in Albino Rat Fetuses." *Anatomical Record* 45:254.

_____. 1935. "Further Studies Upon Development of Somatic Activities in Albino Rat Fetuses." *Proceedings of New York Society of Experimental Biology* 32:621–622.

Aries, P. 1965. *Centuries of Childhood.* New York: Vintage Books.

Bayley, N. 1939. "Mental and Motor Development From Two to Twelve Years." *Review of Educational Research* 9:18–37.

Bean, R.B. 1924. "The Pulse of Growth in Man." *Anatomical Record* 28:45-61.

Bertalanffy, L. Von. 1933. *Modern Theories of Development.* New York: Oxford University Press.

Bolaffio, M. and G. Artom. 1924. "Ricerche Sulla Fisiologia del Sistem Nervosa del feto Umano." *Archiva de Scientifica Biologia* 5:457–487.

Bolton, J. and J.M. Moyes. 1912. "The Major Cyto-Architecture of the Cerebral Cortex of the Human Fetus of Eighteen Weeks." *Brain* 35:1–25.

Boring, E.G., H.S. Langfeld, and H.P. Weld. 1939. *Introduction to Psychology.* New York: Wiley.

Bowlby, J. 1969. *Attachment and Loss. Vol. 1: Attachment.* New York: Basic Books.

Breed, F.S. 1911. "The Development of Certain Instincts and Habits in Chicks." *Behavior Monographs* 1:1–178.

Bridgman, C.S., and L. Carmichael. 1935. "An Experimental Study of the Onset of the Behavior in the Fetal Guinea Pig." *Journal of Genetic Psychology* 47:247–267.

Buhler, C. 1930. *The First Year of Life.* New York: Day. Reprint Arno, 1975

Burr, H.S. and C.L. Hovland. 1937. "Bioelectric Correlates of Development in Amblystoma." *Yale Journal of Biological Medicine* 9:541–545.

Campbell, R.V.D. and A.A. Weech. 1941. "Measures Which Characterize the Individual During Development of Behavior in Early Life." *Child Development* 12:217–236.

Carmichael, L. 1926. "The Development of Behavior in Vertebrates Experimentally Removed from the Influence of External Stimulation." *Psychological Review* 83:61–68.

285

_____. 1927. "A Further Study of the Development of Behavior in Vertebrates Experimentally Removed from the Influence of External Stimulation." *Psychological Review* 34:34–47.

_____. 1928. "A Further Study of the Development of Behavior in Vertebrates Experimentally Removed from the Influence of External Stimulation." *Psychological Review* 35:253–260.

_____. 1933. "Origin of Prenatal Growth of Behavior." In *A Handbook of Child Psychology.* 2nd. rev. ed, ed. C. Murchison. Worcester, MA: Clark University Press.

_____. 1936. "A Re-evaluation of the Concepts of Maturation and Learning as Applied to the Early Development of Behavior." *Psychological Review* 43:450–470.

_____. 1941. "The Experimental Embryology of Mind." *Psychological Bulletin* 38:1–28.

_____. ed. 1946. *Manual of Child Psychology.* New York: Wiley.

Castle, W.E. 1921. *Genetics and Eugenics.* Cambridge: Harvard University Press.

Chaney, L.B. and M.B. McGraw. 1932. "Reflexes and Other Motor Activities in Newborn Infants." *Bulletin of the Neurological Institute of New York* 2:1–56.

Child, C.M. 1924. *Physiological Foundations of Behavior.* New York: Holt.

_____. 1939. "Physiological Gradients in Relation to Development." *Current Science* Special Number: 51–57.

_____. 1940. "Lithium and Echinoderm Exogastrulation." *Physiological Zoology* 13:4–4.

Coghill, G.E. 1929a. "The Early Development of Behavior in Amblystoma and in Man." *Archives of Neurology and Psychiatry* 21:989–1009.

_____. 1929b. *Anatomy and the Problem of Behavior.* New York: Cambridge University Press.

_____. 1930a. "Individuation versus Integration in the Development of Behavior." *Journal of Genetic Psychology* 3:431–435.

_____. 1930b. "Structural Basis for Integration of Behavior." *Proceedings of the National Academy of Science* 16:637–643.

_____. 1933. "The Neuro-Embryologic Study of Behavior: Principles, Perspectives, and Aims." Science 78:131–138.

_____. 1940. "Early Embryonic Somatic Movements in Birds and in Mammals other than Man." *Child Development Monographs* 5:1–48.

Conel, L.J. 1939a. *The Postnatal Development of the Human Cerebral Cortex: Vol. 1 Cortex of the Newborn.* Cambridge: Harvard University Press.

_____. 1939b. "The Brain Structure of the Newborn Infant and Consideration of the Senile Brain." *Research Publications for Association of Nervous and Mental Disorders* 19:247–255.

_____. 1941. *The Postnatal Development of the Human Cerebral Cortex: Vol. 2 Cortex of the One Month Infant.* Cambridge: Harvard University Press.

Conklin, E.G. 1916. *Heredity and Environment in the Development of Man.* Princeton: Princeton University Press.

Coronios, J.D. 1933. "Development of Behavior in the Fetal Cat." *Genetic Psychology Monographs* 14:283–286.

Courtis, S.A. 1935. "Maturation as a Factor in Diagnosis." *Yearbook of the National Society of Student Education* 34:169–187.

Crowell, D.H. 1967. "Infant Motor Development." In *Infancy and Early Childhood*, ed. Y. Brackbill. New York: Free Press.

Danforth, C.H. 1932. "Artificial and Hereditary Suppression of Sacral Vertebrae in the Fowl." *Proceedings of the Society for Experimental Biology* 30:437–438.

Danzinger, L. and L. Frankl. 1934. "Zum Problem der Funktionsreifund." *Z. Kinderforsch* 43:19–254.

Dashiell, J.F. 1928. *Fundamentals of Objective Psychology.* Boston: Houghton Mifflin.

Darwin, C. 1877. "Biographical Sketch of an Infant." *Mind* 2:285–294.

De Crinis, M. 1932. "Die Entwicklung der Grosshirnrinde nach der Geburt in Ihren Beziehungen zur Intellektuellen Ausreifung des Kindes." *Wien Klin. Wschr* 45:1161–1165.

Dennis, W. 1934a. "Congenital Cataract and Unlearned Behavior." *Journal of Genetic Psychology* 44:30–351.

———. 1934b. "A Description and Classification of the Responses of the Newborn Infant." *Psychological Bulletin* 31:5–22.

———. 1935. "The Effect of Restricted Practice upon the Reaching, Sitting, and Standing of Two Infants." *Journal of Genetic Psychology* 47:17–32.

———. 1938. "Infant Development Under Conditions of Restricted Practice and of Minimum Social Stimulation: A Preliminary Report." *Journal of Genetic Psychology* 53:149–158.I

———. 1940. "Does Culture Appreciably Affect Patterns of Infant Behavior?" *Journal of Social Psychology* 12:305–317.

———. 1941. "Spalding's Experiment on the Flight of Birds Repeated with Another Species." *Journal of Comparative Psychology* 31:337–348.

Detwiler, S.R. 1920. "Experiments on the Transplantation of Limbs in Amblystoma." *Journal of Experimental Zoology* 31:117–120.

———. 1928. "Further Experiments upon the Alternation of the Direction of Growth in Amphibian Spinal Nerves." *Procedings of the Journal of Experimental Society* 51:8

———. 1931. "Problems in the Development of the Nervous System." *Journal of Nervous and Mental Disease* 73.

Dewey, E. 1935. *Behavior Development in Infants: A Survey of the Literature on Prenatal and Postnatal Activity, 1920–1934.* New York: Columbia University Press. Reprint edition New York: Arno Press, 1975.

Dewey J. 1934. *Art as Experience.* New York: Minton, Balch and Company.

———. 1938. *Logic: The Theory of Inquiry.* New York: Henry Holt.

Dobzansky, T. 1936. In *Biological Effects of Radiation: Vol. 2,* ed. B.M. Duggar, New York: McGraw Hill.

Dunn, L.C. 1939–1940. "Heredity and the Development of Early Abnormalities in Vertebrates." In *Harvey Lectures* New York Academy of Medicine: Lancaster Science Press.

Durken, B. 1932. *Experimental Analysis of Development.* New York: Norton.

Elkind, D. 1986. "Formal Education and Early Childhood Education: An Essential Difference." *Phi Delta Kappan* (May):631–636.

Endler, N.S., L.R. Boulter, and H. Osser, eds. 1968. *Contemporary Issues in Developmental Psychology.* New York: Holt, Rinehart and Winston.

Fenton, J. 1925. *A Practical Psychology of Babyhood*. Boston: Houghton Mifflin.

Fowler, W. 1962. "Teaching a Two-Year-Old to Read." *Genetic Psychology Monographs* 66:181–283.

Gates, A.I. 1930. *Psychology for Students of Education*. Rev. ed. New York: Macmillan

Gesell, A. 1925. *The Mental Growth of the Pre-School Child: A Psychological Outline of Normal Development from Birth to the Sixth Year, Including a System of Developmental Diagnosis*. New York: Macmillan.

_____. 1928. *Infancy and Human Growth*. New York: Macmillan.

_____. 1929. "Maturation and Infant Behavior Pattern." *Psychological Review* 36:307–319.

Gesell, A. and H. Thompson. 1929. "Learning and Growth in Identical Infant Twins: An Experimental Study of the Method of Co-Twin Control." *Genetic Psychology Monographs* 6:1–124.

_____. 1933. "Maturation and the Patterning of Behavior." In. *A Handbook of Child Psychology*, 2nd ed. rev., ed. C. Murchison. Worcester: Clark University Press

Gesell, A., H. Thompson, and C.S. Amatruda. 1934. *Infant Behavior: Its Genesis and Growth*. New York: McGraw-Hill.

Gilchrist, F.G. 1933. "The Time Relations of Determinations in Early Amphibian Development." *Journal of Experimental Zoology* 66:15–49.

Goldschmidt, R. 1938. *Physiological Genetics*. New York: McGraw Hill.

Goodenough, F.L.; K.M. Maurer, and M.J. Van Wagenen. 1940. *Minnesota Preschool Scale*. Minneapolis: Educational Test Bureau.

Gottlieb, G. 1976. "Conceptions of Prenatal Development." In *Behavioral Embryology*, ed. G. Gottlieb. New York: Academic Press.

Halverson, H.M. 1931. "An Experimental Study of Prehension in Infants by Means of Systematic Cinema Records." *Genetic Psychology Monographs* 10:107–286.

_____. 1932. "A Further Study of Grasping." *Journal of Genetic Psychology* 7:34–64.

Hilgard, J.R. 1932. "Learning and Maturation in Preschool Children." *Journal of Genetic Psychology* 31:36–56.

_____. 1933. "The Effect of Early and Delayed Practice on Memory and Motor Performances Studied by the Method of Co-Twin Control." *Genetic Psychology Monographs* 14:493–567.

Holmes, J.S. 1922. "A Tentative Classification of the Forms of Animal Behavior." *Journal of Comparative Psychology* 2:173–186.

Hooker, D. 1936. "Early Fetal Activity in Mammals." *Yale Journal of Biological Medicine* 8:579–602.

_____. 1937. "The Development of Reflexes in the Mammalian Fetus." *Anatomical Record* (Supplement) 70:55.

_____. 1939. "Fetal Behavior." *Research Publications of the Association of Nervous and Mental Disorders* 19:237–243.

Huxley, J. 1931. *Problems of Relative Growth*. New York: Dial Press.

Irwin. O.C. 1932. Organismic Hypothesis and Differentiation of Behavior." *Psychological Review* 39:128–146.

Jennings, H.S. 1935. *Genetics.* New York: Norton.

Jerslid, A.T. 1932. "Training and Growth in the Development of Children: A Study of the Relative Influence of Learning and Maturation." *Child Development Monographs* 10:78.

Koffka, K. 1924. *The Growth of the Mind.* Trans. R.M. Ogden. London: Kegan Paul.

Krabbe, K. 1912. "Les Reflexes Chez le Foetus." *Review of Neurology* 24:434–435.

Kuo, Z.Y. 1930. "The Genesis of Cats' Responses to the Rat." *Journal of Comparative Psychology* 11:1–35.

———. 1939. "Total Pattern or Local Reflexes?" *Psychological Review* 46:93–122.

Lamarck, J.B. 1773. *Philosophie Zoologique.* Paris.

Langworthy, O.R. 1930. "Medullated Tracts in the Brain Stem of a Seven-Month Human Fetus." *Carnegie Institution of Washington: Contributions to Embryology* 21:37–52.

———. 1933. "Development of Behavior Patterns and Myelinization of the Nervous System in the Human Fetus and Infant." *Carnegie Institution of Washington: Contributions to Embryology* 24:189.

Lashley, K.S. 1930. Basic Neural Mechanisms in Behavior." *Psychological Review* 97:1–24.

Lock, R.H. 1916. *Recent Progress in the Study of Innate Behavior.* New York: Dutton.

Lewin, K. 1936. *Principles of Topological Psychology.* New York: McGraw-Hill.

Lorenz, K. 1935. "Der Kumpan in der Umwelt des Vogels. Der Artgenosse Alsauslosendes Moment Sozialer Verhaltungsweisein." *Journal of Ornithology* 83:137–213.

———. 1973. *Motivation of Human and Animal Behavior: An Ethological View.* New York: Von Nostrand Reinhold.

Marais, E. 1937. *The Soul of the White Ant.* New York: Dodd, Mead, and Company.

Marquis, D.G. 1930. "The Criterion of Innate Behavior." *Psychological Review* 37:334–349.

Matthews, S.A. and S.R. Detwiler. 1926. "The Reactions of Amblystoma Embryos Following Prolonged Treatment with Chloretone." *Journal of Experimental Zoology* 45:279–292.

McDougall, W. 1914. *An Introduction to Social Psychology.* Boston: Luce.

McGraw, M.B. 1935. *Growth: A Study of Johnny and Jimmy.* New York: Appleton-Century.

———. 1937. "The Moro Reflex." *American Journal of the Diseases of Children* 54:240–251.

———. 1939a. "Behavior of the Newborn Infant and Early Neuromuscular Development." *Research Publications of the Association of Nervous and Mental Disorders* 19:244–246.

———. 1939b. "Later Development of Children Specially Trained During Infancy." *Child Development* 10:1–19.

———. 1939c. "Swimming Behavior of the Human Infant." *Journal of Pediatrics* 15:485–490.

———. 1940a. "Basic Concepts and Procedures in a Study of Behavior Development." *Psychological Review* 47:79–89.

_____. 1940b. "Neural Maturation as Exemplified in Achievement of Bladder Control." *Journal of Pediatrics* 16:580–590.

_____. 1940c. "Suspension Grasp Behavior of the Human Infant." *American Journal of the Diseases of Children* 60:799–811.

_____. 1940d. "Neuromuscular Mechanism of the Infant: Development Reflected by Postural Adjustments to an Inverted Position." *American Journal of the Diseases of Children* 60:1031–1042.

_____. 1940e. "Neuromuscular Development of the Human Infant as Exemplified in the Achievement of Erect Locomotion." *Journal of Pediatrics* 17:747–771.

_____. 1941a. "Development of Neuromuscular Mechanisms as Reflected in the Crawling and Creeping Behavior of the Human Infant." *Journal of Genetic Psychology* 58:83–111.

_____. 1941b. "Neuromotor Maturation of Anti-Gravity Functions as Reflected in The Development of A Sitting Posture." *Journal of Genetic Psychology* 59:155–175.

_____. 1942. *The Neuromuscular Maturation of the Human Infant.* New York: Columbia University Press.

_____. 1946. "Maturation of Behavior." In *Manual of Child Psychology,* ed. L. Carmichael. New York: Wiley.

_____. 1990. "Memories, Deliberate Recall, and Speculation." *American Psychologist* 45:934–937.

McGraw, M.B. and A.P. Weinbach. 1936. "Quantitative Measures in Studying Development of Behavior Patterns (Erect Locomotion)." *Bulletin of the Neurological Institute of New York* 4:563–571.

Metfessel, M. 1940. "Relationships in Heredity and Environment in Behavior." *Journal of Psychology* 10:177–198.

Minkowski, M. 1922. "Uber fruhzeitige Bewegungen. Reflexe und Muskulare Reaktionen beim Menschlichen Fotus und Ihre Beziehungen zum Fotalen Nerven und Muskelsystem." *Schweiz Medischinishe Yahrbuch* 52:721–724; 751–755.

Moore, O.K. 1960. *Automated Responsive Environments.* Hamden, Ct.: Basic Education, Inc. (a film).

Morgan, T.H. 1932. *The Scientific Basis of Evolution.* New York: Norton.

Mowrer, O.H. 1936. " 'Maturation ' vs. 'Learning' in the Development of Vestibular and Optokinetic Nystagmus." *Journal of Genetic Psychology* 48:383–404.

Munn, N.L. 1938. *Psychological Development: An Introduction to Genetic Psychology.* Boston: Houghton Mifflin.

Northrop, F.S.C. and H.S. Burr. 1937. "Experimental Findings Concerning the Electrodynamic Theory of Life and an Analysis of Their Physical Meaning." *Growth* 1:78–88.

Olson, W.C. 1930. *Problem Tendencies in Children.* New York: Dutton.

Palmer, C.E. 1937. "The Analysis of Longitudinal Data." Unpublished Paper presented at the Eastern Regional Conference of the Society for Research in Child Development, New York City, December 4.

Pankratz, D.S. 1931. "A Preliminary Report on the Fetal Movements in the Rabbit." *Anatomical Record* 48:58–59.

Pavlov, I.P. 1927. *Conditioned Reflexes: An Investigation of the Physiological Activity of the Cerebral Cortex.* Trans. G.V. Anrep, London: Oxford University Press.

Perrin, F.A.C. and D.B. Klein. 1926. *Psychology: Its Methods and Principles.* New York: Holt.

Pratt, K.C. 1933. "The Neonate." In *A Handbook of Child Psychology,* 2nd ed. rev., ed. L. Carmichael. Worcester: Clark University Press.

Pratt, K.C.; A.K. Nelson, and K.H. Sun. 1930. "The Behavior of the Newborn Infant." *Ohio State University Contributions to Psychology* 10:ix–235.

Preyer, W. 1888. *The Mind of the Child: Part 1, The Senses and the Will.* Trans. H. W. Brown. New York: Appleton Century.

Reynolds, M.M. 1928. "Negativism of Pre-School Children: An Observational and Experimental Study." *Contributions to Education* 228. New York: Teacher's College, Columbia University.

Ritter, W.E. and E.W. Bailey. 1927. "The Organismal Conception." *University of California Publications in Zoology* 31:307–358.

Shepard, J.F. and F.S. Breed. 1913. "Maturation and Use in the Development of Instinct." *Journal of Animal Behavior* 3:274–285.

Shinn, M.W. 1900. *Biography of a Baby.* Boston: Houghton Mifflin.

Shirley, M.M. 1931a. *The First Two Years, A Study of Twenty-Five Babies: Vol. 1 Postural and Locomotor Development.* Institute of Child Welfare Monograph Series no. 6. Minneapolis: University of Minnesota Press.

_____. 1931b. "The Sequential Method for the Study of Maturing Behavior Patterns." *Psychological Review* 38:507–528.

_____. 1931c. "Is Development Saltatory as Well as Continuous?" *Psychological Bulletin* 28:664–665.

_____. 1933a. *The First Two Years, A Study of Twenty-Five Babies: Vol. 2 Postural and Locomotor Development.* Institute of Child Welfare Monograph Series no. 7 Minneapolis: University of Minnesota Press.

_____. 1933b. *The First Two Years, A Study of Twenty-Five Babies:Vol. 3, Personality Manifestations.* Institute of Child Welfare Monograph Series no. 7. Minneapolis: University of Minnesota Press.

_____. 1933c. "Locomotor and Visual Manual Functions in the First Two Years," In *A Handbook of Child Psychology,* 2nd ed. rev., ed. L. Carmichael. Worcester: Clark University Press.

Sinnott, E.W. 1937a. "The Genetic Control of Developmental Relationships." *American Naturalist* 71:113–119.

_____. 1937b. "The Relation of Gene to Character in Quantitative Inheritance." *Procedings of the National Academy of Science* 23:224–227.

_____. 1939. "A Developmental Analysis of the Relations Between Cell Size and Fruit Size in Cucurbits." *American Journal of Botany* 26:179–189.

Sinnott, E.W. and L.C. Dunn. 1939. *Principles of Genetics.* New York: McGraw-Hill.

Spalding, D.A. 1873. "Instinct: With Original Observations on Young Animals." *Macmillian's Magazine* 27:282–293.

_____. 1875. "Instinct and Acquisition." *Nature* 12:507–508.

Spemann, H. 1925. "Some Factors of Animal Development." *British Journal of Experimental Biology* 2:493.

Stadler, L.J. 1939. "The Experimental Alteration of Heredity." *Growth* 3:19–36.

Stern, C. 1939. "Recent Work on the Relation Between Genes and Developmental Processes." *Growth* (Supplement: First Symposium on Development and Growth) 3:19–36.

Strassmann, P. 1903. "Das Leben vor der Geburt." *Samml. Klin. Vortr, N. F., Gyndk* 353:947–968.

Strayer, L.C. 1930. "Language and Growth: The Relative Efficacy of Early and Deferred Vocabulary Training Studied by the Method of Co-Twin Control." *Genetic Psychology Monographs* 8:209–319.

Stutsman, R. 1931. *Mental Measurement of Preschool Children.* Yonkers-on-Hudson: World Book.

Sunley, R. 1968. "Early Nineteenth Century Literature on Child Rearing." In *Readings in Behavior and Development,* ed. D E. Ellis. New York: Holt, Rinehart, and Winston.

Swenson, E.A. 1929. "The Active Simple Movements of the Albino Rat Fetus: The Order of Their Appearance, Their Qualities, and Their Significance." *Anatomical Record* 42:40.

Thorndike, E.L. 1919. *Educational Psychology: Vol. 1, The Original Nature of Man.* New York: Teachers College.

Tilney, F. 1933. "Behavior in its Relation to the Development of the Brain: II. Correlation Between the Development of the Brain and Behavior in the Albino Rat from Embryonic States to Maturity." *Bulletin of the Neurological Institute of New York* 3:252–358.

———. 1937. "The Structure and Development of the Brain." Unpublished Lecture, New York.

Tilney, F. and L.S. Kubie. 1931. "Behavior in its Relation to the Development of the Brain." *Bulletin of the Neurological Institute of New York* 3:229–313.

Tracy, H.C. 1926. "The Development of Motility and Behavior Reactions in the Toadfish (Opanus Tau)" *Journal of Comparative Neurology* 66:157–175.

Tuge, H. 1931. "Early Behavior of Embryos of the Turtle, Terrapene Carolina." *Procedings of the Society of Experimental Biology of New York* 29:52–53.

———. 1937. "The Development of Behavior in Avian Embryos." *Journal of Comparative Neurology* 66:157–175.

Waddington, C.H. 1936. *How Animals Develop.* London: Norton.

———. 1939. *An Introduction to Modern Genetics.* London: Allen and Unwin.

Watson, J.B. 1914. *Behavior: An Introduction to Comparative Psychology.* New York: Holt.

———. 1919. *Psychology from the Standpoint of a Behaviorist.* Philadelphia: Lippincott.

———. 1928. *The Psychological Care of Infant and Child.* New York: W.W. Norton.

———. 1930. *Behaviorism* Rev., New York: Norton.

Weinbach, A.P. 1937. "Some Physiological Phenomena Fitted to Growth Equations: I. Moro Reflex." *Human Biology* 9:549–555.

Weinbach, A.P. 1940. "Some Physiological Phenomena Fitted to Growth Equations: IV. Time and Power Relations for a Human Infant Climbing Inclines of Various Slopes." *Growth* 4:128–134.

Weismann, A. 1889. *Essays Upon Heredity and Kindred Biological Problems.* Oxford: Clarendon Press.

Weiss, A.P. 1929. *A Theoretical Basis of Human Behavior.* Columbus, Ohio: Adams.

Weiss, P. 1939. *Principles of Development.* New York: Holt.

Windle, W.F. 1933. "The Instinct Hypothesis versus the Maturation Hypothesis." *Psychological Review* 40: 35–59.

———. 1930. "Normal Behavioral Reactions of Kittens Correlated with the Postnatal Development of Nerve-Fiber Density in the Spinal Gray Matter." *Journal of Comparative Neurology* 50:479–503.

Windle, W.F. and A.M. Griffin. 1931. "Observations on Embryonic and Fetal Movements of the Cat." *Journal of Comparative Neurology* 52:149–188.

Windle, W.F., J.E. O'Donnell and E.E. Glasshagle, 1933. "The Early Development of Spontaneous and Reflex Behavior in Cat Embryos and Fetuses." *Physiological Zoology* 6:521–541.

Windle, W.F. and D.W. Orr. 1934. "The Development of Behavior in Chick Embryos: Spinal Cord Structure Correlated with Early Somatic Motility." *Journal of Comparative Neurology* 60:287–308.

Witty, P.A. and H.C. Lehman. 1933. "The Instinct Hypothesis versus the Motivation Hypothesis." *Psychological Review* 40:33–59.

Woodger, J.H. 1937. *The Axiomatic Method in Biology.* New York: Cambridge University Press.

Woodworth, R.S. 1921. *Psychology. A Study of Mental Life.* New York: Holt.

Yanase, J. 1907. "Beitrage zur Physiologie des Peristaltischen Bewegungen des Embryonalen Darmes." *Pflug. Archive ges. Physiologie* 117:345–382; 119:451-464.

Yerkes, R.M. and D. Bloomfield. 1910. "Do Kittens Instinctively Kill Mice?" *Psychological Bulletin* 7:253–263.

About the Book and Editors

Myrtle McGraw's pioneering contributions to the field of child development have been readily acknowledged and documented, yet controversy persists among psychologists as to how to interpret her ideas about significant factors that influence learning. This collection includes some of McGraw's most cogent work, including five previously unpublished essays that address misconceptions and clarify her principles of development. These essays demonstrate that McGraw conceived of development as a continuous interaction between neural and behavioral growth processes that could not be reduced to either heredity or environment.

The editors document McGraw's little-known collaboration in the 1930s with John Dewey and several other notable scientists. Dewey believed their research promised to "revolutionize work in the field of child study." Their collaboration brings to light new evidence that McGraw's work made use of novel methods to study developmental behavior and enabled Dewey to examine the origin and role of judgment in inquiry. Five other contributors discuss specific issues and episodes that illuminate why McGraw's scientific innovations remain pertinent to researchers in infant motor development.

Thomas C. Dalton is senior research associate and lecturer in politics and society at California Polytechnic State University at San Luis Obispo. **Victor W. Bergenn** is executive director of the Council on Educational Psychology in Leonia, New Jersey.

Index

AAAS. *See* American Association for the Advancement of Science
American Association for the Advancement of Science (AAAS), 63
American Psychological Association (APA), 54, 63
APA. *See* American Psychological Association
Attitude, 9, 24–25, 88–91, 93, 132–134, 146, 148, 277
  acquiescent, 84, 134
  negative, 25, 88, 91, 148
Awareness. *See* Brain, consciousness

Baldwin, J. M., 210
Behavior, ix, 163, 170, 177, 193, 216–217, 224–225, 229–237, 240–241, 250–252, 257
  conditioning, 6, 40, 70, 73, 155, 164–166, 177–178, 193, 201, 225–226, 273
  consistency, 208, 213, 227, 264, 269, 277
  course, 171, 184–186, 251
  feeding, 107
  fixity of, 81–82, 88–90, 173, 180–181, 183–185, 236
  modification of, 89–90, 171–172, 184–187, 236–237
  patterns, 171, 173–174, 178–186, 227, 236, 253
  phases of, 140–143, 172, 180, 182–183, 199
  pre-natal, xi, 141, 155, 240–242, 275–276, 280
  repetition, 100, 235–236
  restriction of, 112–115, 172, 208, 231–236, 267
  social, 147–148
  spontaneous, 156, 254, 275–276
  systems, 13–14
  theories of, 13, 21–23, 263
  variability of, 29, 271
  See also Development; Growth; Locomotion; Reflexes
Behaviorism, ix, 20, 29, 51, 110, 164, 212, 224, 226, 280
Bertalanffy, L. von, 211, 217
Biological systems, 228–229
Brain, 1, 3, 9–11, 13, 24, 197–203, 236, 257, 272, 274–276, 279
  and axon growth, 129, 137
  cerebral cortex, 13, 15, 25, 27, 118, 143, 155, 180, 198–200, 204, 248, 252, 275, 279–280
  consciousness, 7, 13–14, 18, 21, 24–25, 30, 34, 118, 143, 186, 204, 248, 252, 275, 279–280
  cortical inhibition, 133–134, 137, 140–141, 147, 159, 166, 237, 252–254, 271, 280
  development, 13, 21–23, 62, 138, 208, 238–241, 246–247
  evolution of, 9, 17, 24–25
  hierarchy, 138, 199, 246, 248, 275
  and intelligence, 8–9, 22–23, 24, 279
  and memory, 25–26, 99–100
  neuronal connections in, 22–23, 129, 137, 190–191
  sub-cortex, 25, 134, 143, 147, 159, 180, 190, 198–199, 252–254, 275, 279–280
  See also Development; Neuroembryology; Neuromuscular system
Briarcliff College, 3, 26
Butler, S., 21

297